HANDBOOK OF CHEMISTRY,

FOR

SCHOOL AND HOME USE.

BY

W. J. ROLFE AND J. A. GILLET,

TEACHERS IN THE HIGH SCHOOL, CAMBRIDGE, MASS.

BOSTON:

WOOLWORTH, AINSWORTH, & CO.

NEW YORK: A. S. BARNES & CO.

1869.

CAMBRIDGE:
PRESS OF JOHN WILSON AND SON.

PREFACE.

We have prepared this book at the urgent request of many teachers who desire something easier and briefer than the "Chemistry" of the Cambridge Course of Physics. This is a more elementary work, and therefore less theoretical. It is, in a word, a manual of the elementary facts and principles of Chemistry. While there is little that is strictly theoretical in the body of the book, we have not forgotten that Chemistry is not a mere mass of dead facts, but a living science. We have therefore aimed to present the facts in such a way as to awaken a love for the science itself, no less than to show its utility. Our object has been not only to describe the properties of the leading elements and compounds, but also to show the wonderful part played by the chemical force in nature, as well as to illustrate its practical applications in the arts. We have endeavored to give the young student a glimpse of—

> ——"the circle of eternal change,
> Which is the life of nature,"—

a life whose fountain-head is the sun, and whose

processes we trace in the chemistry of the atmosphere. In this indirect but natural way, we have introduced as much of *organic* chemistry as could well find place in an elementary manual.

The *metals* we have grouped more according to their uses than according to their chemical relations, with a view to make the subject less dry and more practical.

It is difficult to write a book suited to the wants of different schools, or even of successive classes in the same school; since the schools vary in grade, and the classes in ability. We have attempted to meet this difficulty by putting into the body of the book the simpler and more practical portions of the subject, which can be readily mastered by young pupils of average ability, and by adding in the Appendix a concise statement of the elements of chemical philosophy as now understood. In this way, the simpler and the more difficult parts, instead of being confusedly mixed together, distinguished merely by the common and clumsy expedient of coarse and fine print, are treated separately, so that each may be complete and symmetrical. At the same time, the two parts do not overlap, but form successive chapters progressively arranged. Our experience as teachers has convinced us, that this is, on the whole, the best method of meeting the wants of successive classes; and the marked favor with which our "Handbook of the Stars," an elementary "Astronomy" on precisely the same plan, has been received, shows that the experience of many of our fellow-teachers has been pretty much the same as our own.

We have not forgotten that the theories of chemistry have been revolutionized within a few years; and we have of course used the new symbols and notation, without which it is as impossible to teach modern chemistry as it would be to teach modern astronomy in the language of the old astrologers. Until within a year or two, there was some excuse for clinging to the old notation, as rival systems were contending for the supremacy. But even as long ago as October, 1866, Miller, the first English authority on the subject, wrote as follows: "The change in notation is certain to be adopted, since in none of the recent investigations in this country, and in very few of those on the Continent, is the old method made use of." Since that time, the system of symbols and notation used by us in this book has been formally adopted by the Chemical Society of London, whose decision is recognized as law in England; and it is now taught in Harvard College, which will certainly be acknowledged as one of the highest authorities in this country.

In the chapter on the Chemistry of the Atmosphere, we have been especially indebted to the Lectures on "Religion and Chemistry" (New York, 1866), by Professor Josiah P. Cooke, Jr., of Harvard College; and the chapter on Chemical Philosophy is almost wholly drawn from the "First Principles of Chemical Philosophy" (Cambridge, 1868), by the same eminent author. His remarkable power of presenting difficult subjects in a way at once clear and attractive, makes these books especially valuable to the teacher.

We have been under obligations, also, to Miller's "Elements of Chemistry" (London, 1866), and Roscoe's "Elementary Chemistry" (London, 1866). These books (both of which, we are glad to see, have been recently reprinted here) as well as Hofmann's "Modern Chemistry" (London, 1865), and Wurtz's "Introduction to Chemical Philosophy" (London, 1867), the teacher will find useful for purposes of reference.

CAMBRIDGE, December 10, 1868.

☞ Teachers will observe that, except in the Appendix, either the old chemical name or the ordinary commercial name of the substance is generally added, in parentheses, to the new chemical name.

TABLE OF CONTENTS.

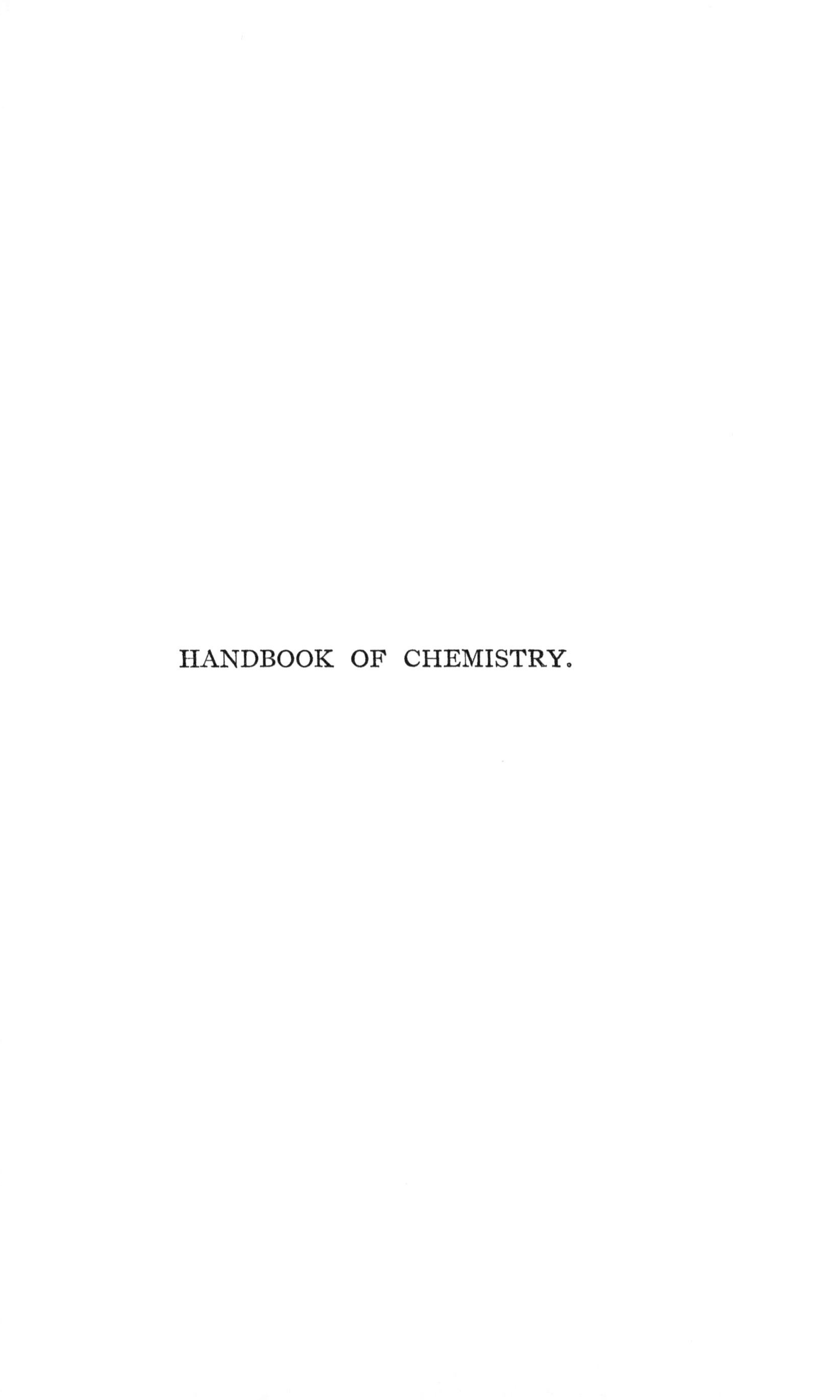

HANDBOOK OF CHEMISTRY.

THE NON-METALLIC ELEMENTS.

COMPOUNDS AND ELEMENTS.

1. *Muriatic Acid.* — Into a bottle arranged as shown in Figure 1, put a few bits of zinc, and pour a little muriatic acid over them. Effervescence at once takes place, and a colorless gas passes through the bent tube into the jar, which has been previously filled with water, and in-

Fig. 1.

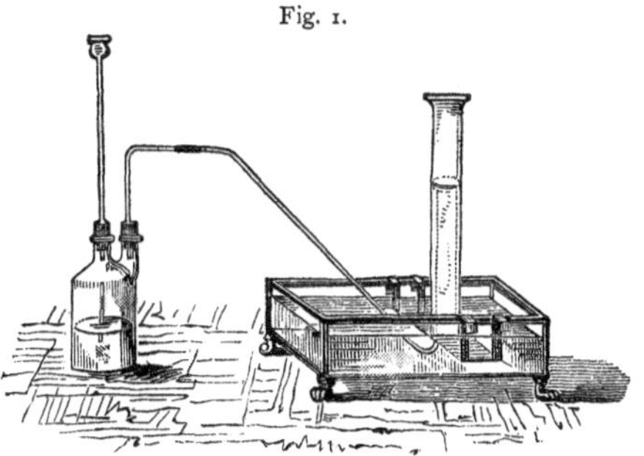

verted over the trough. When the jar is full of gas, raise it from the water, and apply a lighted taper to its mouth. The gas takes fire with a slight explosion, and burns with a pale flame. This gas is called *hydrogen*, and comes from the muriatic acid.

Pour a little of the muriatic acid into a flask (Figure 2), add a little black oxide of manganese, mix them thoroughly, and heat gently. A greenish-yellow gas passes into the jar inverted over the water-trough. When the jar is full, close its mouth with a glass plate, remove it from the trough, and set it on the table. Moisten a piece of litmus-paper or colored cotton cloth, and hold it inside the jar. It will be instantly bleached. This gas is called *chlorine*, and, like the hydrogen, comes from the muri-

Fig. 2.

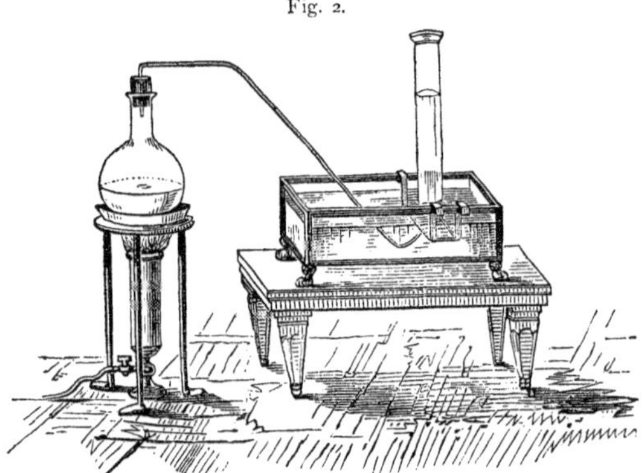

atic acid. Chlorine may be recognized by its bleaching power, and also by its color and its suffocating odor.

We have now got from muriatic acid two gases, hydrogen and chlorine.

If a small glass jar be filled with equal parts of these two gases, and a light be applied to its mouth, the mixture will take fire with an explosion. If a little dish of ammonia be standing by, a dense white cloud will appear when the gases burn.

Now this cloud shows that muriatic acid has been produced; for, if we dip a glass rod in muriatic acid, and hold it over the ammonia, the white cloud at once appears.

Muriatic acid, then, contains nothing but hydrogen and chlorine.

2. *Water.*—If a bit of sodium be thrust beneath the mouth of an inverted jar, filled with water, as shown in Figure 3, bubbles of gas rise and fill the jar. On raising the jar, and applying a taper to its mouth, we see, by the way in which the gas takes fire, that it is hydrogen.

Fill a tall glass jar with *chlorine water* (that is, water which has absorbed chlorine gas), invert it, with its

Fig. 3.

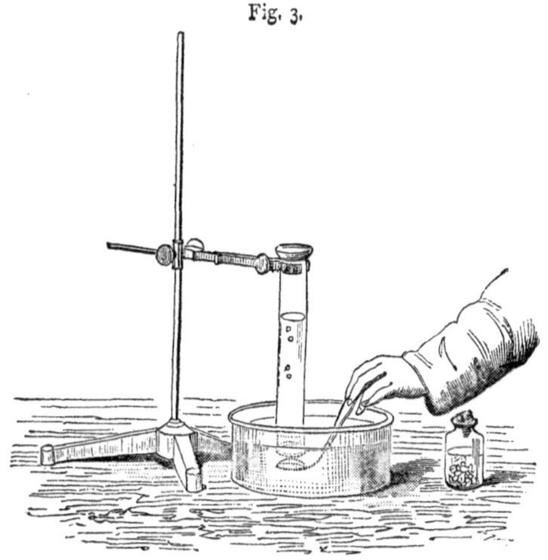

mouth in a shallow dish of water, and leave it for ten or twelve hours in the sunshine. Bubbles of gas will be seen to rise, and collect in the top of the jar. If now we set the jar upright, and plunge a lighted taper into the gas, the taper burns much more brightly than in the air. This gas is called *oxygen.*

We have thus got hydrogen and oxygen from water. If, now, we burn a jet of hydrogen in a jar of oxygen, moisture will collect on the sides. This moisture is water, and comes from the union of the hydrogen and the oxygen. This shows that there is nothing but oxygen and hydrogen in water.

3. *Ammonia.* — Fill with chlorine a long glass tube, closed at one end. Close the other end with a cork, through which passes a dropping tube. (Figure 4.) If, now, a little ammonia be dropped into the tube, flashes of light and dense white fumes appear within. Next allow water to run into the tube as long as it will, and test the gas which remains, by putting a piece of litmus-paper into it. The paper is not bleached, hence the gas is not chlorine. Plunge a lighted taper into it, and it goes out, and the gas does not take fire; hence the gas is neither oxygen nor hydrogen. It is *nitrogen*, and it always, as in this case, extinguishes a lighted taper.

Fig. 4.

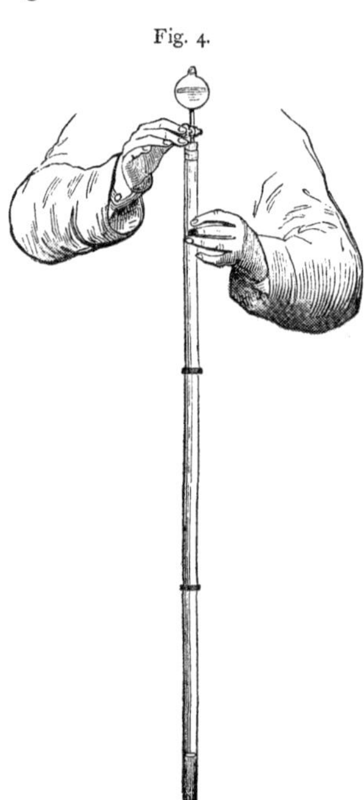

This nitrogen must have come from the ammonia, since there was no nitrogen in the water poured into the tube.

The white fumes which appeared in the tube, when the ammonia was dropped in, showed that muriatic acid was formed. Now we have learned that muriatic acid is made up of chlorine and hydrogen. The hydrogen must also have come from the ammonia.

Ammonia, then, contains nitrogen and hydrogen. It has been found by experiment that it contains nothing but these two gases.

4. *Compounds and Elements.* — We have now seen that muriatic acid, water, and ammonia, are made up of

other substances, which we can separate from them. In other words, they are *compound* substances, and we have *decomposed* them. But no means has been found of decomposing oxygen, hydrogen, chlorine, and nitrogen into simpler substances. These, and other substances which cannot be decomposed, are called *elements*.

Most substances with which we are familiar are compounds. The metals, however, are all elements.

There are only about sixty-five elements in all, and many of these are very rare substances.

5. *Names and Symbols of Elements.* — Elements have been divided into two classes, *metallic* and *non-metallic*. The following list contains the names of the most important elements usually put in each class, together with the abbreviations of these names commonly used. These abbreviations are called *symbols*.

NON-METALLIC ELEMENTS.

Name.	Symbol.	Atomic Weight.	Name.	Symbol.	Atomic Weight.
Boron	B	11	Iodine	I	127
Bromine	Br	80	Nitrogen	N	14
Carbon	C	12	Oxygen	O	16
Chlorine	Cl	35.5	Phosphorus	P	31
Fluorine	F	19	Silicon	Si	28
Hydrogen	H	1	Sulphur	S	32

METALLIC ELEMENTS.

Name.	Symbol.	Atomic Weight.	Name.	Symbol.	Atomic Weight.
Aluminium	Al	27.4	Magnesium	Mg	24
Antimony (*stibium*)	Sb	122	Manganese	Mn	55
Arsenic	As	75	Mercury (*hydrargyrum*)	Hg	200
Barium	Ba	137	Nickel	Ni	58.7
Bismuth	Bi	210	Platinum	Pt	197.5
Calcium	Ca	40	Potassium (*kalium*)	K	39
Chromium	Cr	52.5	Silver (*argentum*)	Ag	108
Cobalt	Co	58.7	Sodium (*natrium*)	Na	23
Copper (*cuprum*)	Cu	63.5	Strontium	Sr	87.5
Gold (*aurum*)	Au	197	Tin (*stannum*)	Sn	118
Iron (*ferrum*)	Fe	56	Zinc	Zn	65.2
Lead (*plumbum*)	Pb	207			

It will be noticed that in some cases the symbol is an abbreviation of the *Latin* name of the element.

6. *Affinity.* — The force which draws elements together, and locks them up in compounds, is called *affinity*, or *chemical force.*

One of the characteristics of this force is, that *it changes the properties of the substances which it brings together.* It not only unites different substances, but unites them so as to form a new substance. It *combines* them.

This characteristic is well illustrated in the case of water, a substance wholly unlike hydrogen, which is combustible, and oxygen, which aids combustion.

The second characteristic of affinity is, that *it always acts between definite quantities of matter.* This may be illustrated by the following experiment: —

Partially fill the closed arm of the U tube (Figure 5) with a mixture of hydrogen and oxygen, using more than twice as much hydrogen as oxygen. Explode the mixture by means of an electric spark from a Leyden jar. The hydrogen and the oxygen combine to form water, but some gas remains, which, if tested with a taper, is found to be hydrogen.

Fig. 5.

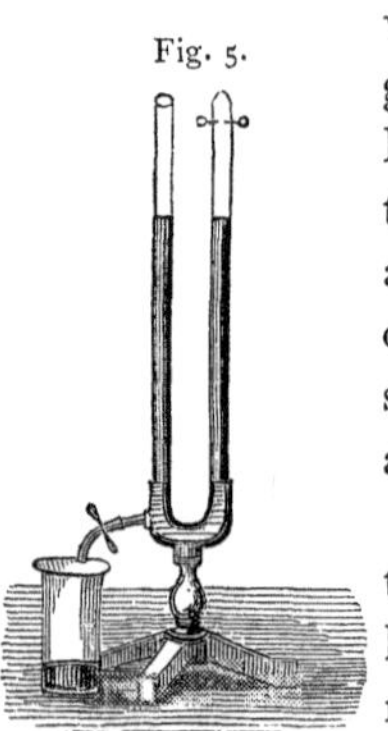

If now we put into the tube a mixture containing *less* than twice as much hydrogen as oxygen, the gas which remains after the explosion is found to be oxygen.

In both cases, the oxygen combines with exactly twice its bulk of hydrogen, and this will be true in whatever proportions the gases may be mixed.

This experiment also illustrates the *third* characteristic of affinity, which is, that *it is often dormant until it is roused to activity by heat, or some other force.* The oxygen and the hydrogen showed no disposition to combine until the spark was sent through them.

7. *Molecules and Atoms.* — It is supposed that all bodies are made up of small particles called *molecules*, and that these molecules are made up of smaller particles, which are called *atoms* (that is, *not to be divided*), since we are not able to subdivide them.

In an elementary substance, the atoms are supposed to be all alike in form and weight, but those of one element differ in weight from those of another. In a *compound* substance, the *molecules* are supposed to be alike, each being made up of a fixed number of atoms of each of the component elements; but the molecules of one compound differ from those of another in their atomic constitution. This theory of the composition of substances is called the *atomic theory.*

Thus the molecules of *muriatic acid* are supposed to be made up of *one* atom of hydrogen and *one* of chlorine; the molecules of *water* to be made up of *two* atoms of hydrogen and *one* of oxygen; and the molecules of *ammonia* to be made up of *three* atoms of hydrogen and *one* of nitrogen.

8. *Symbols of Compounds.* — The symbol of a compound is made up of the symbols of its elements. It also indicates the atomic constitution of the molecules of the compound. The symbol of each element represents *one atom* of that element. If there is more than one atom of the element in the molecule, this is indicated by a small figure placed at the right of the symbol. Thus the symbol for *muriatic acid* is HCl; indicating that muriatic acid is a compound of hydrogen and chlorine, and that a molecule of muriatic acid is made up of one atom of hydrogen and one of chlorine. The symbol for *water* is H_2O, since water is a compound, whose molecule contains two atoms of hydrogen and one of oxygen. The

symbol of *ammonia* is H_3N, its molecule being made up of three atoms of hydrogen and one of nitrogen.

9. *Atomic Weights.* — According to the atomic theory, an atom of chlorine weighs 35.5 times as much as an atom of hydrogen; an atom of oxygen weighs 16 times as much as one of hydrogen; and an atom of nitrogen 14 times as much as one of hydrogen. If then we represent the weight of an atom of hydrogen by 1, the weight of an atom of chlorine will be 35.5; that of an atom of oxygen, 16; and that of an atom of nitrogen, 14. These numbers are called the *atomic weights* of these elements. (5.)

We now see that the symbol of a compound represents not merely the atomic constitution of the compound, but also the relative weights of its elements. Thus *muriatic acid*, HCl, contains 1 part by weight of hydrogen to 35.5 parts by weight of chlorine; *water*, H_2O, contains 2 parts by weight of hydrogen to 16 of oxygen; and *ammonia*, H_3N, contains 3 parts by weight of hydrogen to 14 of nitrogen.

OXYGEN.

10. *Preparation of Oxygen.* — Oxygen is most readily obtained from potassic chlorate (chlorate of potash), a substance of whose weight it forms about 40 per cent. The chlorate is heated in a flask (Figure 6), and the gas is

Fig. 6.

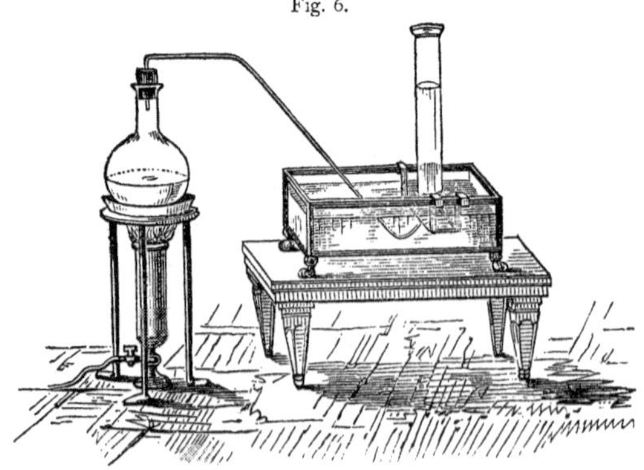

collected over water. Potassic chlorate is a compound of chlorine, potassium, and oxygen, and its symbol is $KClO_3$. The oxygen is set free by the heat, leaving the chlorine and potassium as a compound called *potassic chloride.*

A chemical change of this kind is called a *reaction*, and may be expressed in the form of an equation. In this case the equation will be —

$$KClO_3 = KCl + O_3$$

indicating that the potassic chlorate, $KClO_3$, has been broken up into potassic chloride, KCl, and oxygen, O.

If a small quantity of black oxide of manganese be mixed with the potassic chlorate, the gas is given off at a much lower temperature, and more gradually. The manganese, however, undergoes no change in the process.

11. *Properties of Oxygen.* — We have already seen (2) that a taper burns much more vividly in oxygen than in the air.

If a piece of charcoal be lighted and plunged into a jar of oxygen, it burns with a very brilliant white light. Pour now some clear lime-water into the jar, and shake it. The liquid becomes milky white, showing that the gas has undergone a change; for oxygen has no effect upon lime-water. Charcoal is a form of the element *carbon*, and the oxygen has combined with it. The compound produced is called *carbonic anhydride* (*carbonic acid*), and its composition is represented by the symbol, CO_2.

If a bit of sulphur be ignited and put into a jar of oxygen, it burns with a vivid blue light and a suffocating odor. The sulphur and the oxygen have combined to form *sulphurous anhydride* (*sulphurous acid*), the symbol of which is SO_2.

Phosphorus burns in oxygen with dazzling brilliancy, producing dense, white fumes of *phosphoric anhydride* (*phosphoric acid*), P_2O_5.

Thus far, we have found that substances which burn in the air burn more rapidly and more brightly in oxygen. If, now, we fasten a bit of wood to a steel watch-spring, ignite the wood, and plunge the whole into a jar of oxygen, the watch-spring takes fire, and burns as readily as a splint of wood burns in the air. In this case the iron and the oxygen combine, forming a mixture of two oxides of iron, FeO, *ferrous oxide*, and Fe_2O_3, *ferric oxide.*

The above experiments illustrate several properties of oxygen: (1) the *passive state in which it exists under ordinary circumstances*, it being necessary to *heat* it in order to make it combine rapidly with any substance; (2) the *intense energy with which it enters into combination when once aroused;* and (3) the *wide range of its affinities.* It combines with every known element, except fluorine.

12. *Ozone.*—If a series of electric sparks be sent through dry oxygen, the gas undergoes a change. It acquires more active properties and a peculiar odor from which it takes its name, *ozone.** The odor perceived when an electrical machine is worked is owing to the presence of ozone.

If a paper moistened with a solution of *potassic iodide* (*iodide of potassium*), KI, and starch paste be held opposite a point on the conductor of an electrical machine, the paper becomes blue. The iodine is set free by the ozone, and unites with the starch, forming a blue compound, which colors the paper. Ordinary oxygen has no such power to separate iodine from its combination with potassium.

There has been much discussion concerning the nature of ozone. It appears, however, to be simply oxygen in a modified and condensed form.

* *Ozone* is derived from a Greek word which means *to emit a bad odor.*

This *partially* active state of oxygen is intermediate between the active and passive states already mentioned.

13. *Allotropic States.* — There are other elements which exist in states in which their properties are as unlike as those of oxygen and ozone. These different states of an element are called *allotropic* states.

14. *Oxides.* — "The simple compounds of the elements with oxygen are called oxides, and the specific names of the different oxides are formed by placing before the word *oxide* the name of the element, but changing the termination into *ic* or *ous*, to indicate different degrees of oxidation, and using the Latin name of the element in preference to the English, both for the sake of euphony and in order to secure more general agreement among different languages. When the same element unites with oxygen in more than two proportions, the Latin prepositions or numeral adverbs, *sub*, *per*, *bis*, etc., are prefixed to the word *oxide* in order to indicate the additional degrees. Formerly these compounds were called the oxides of the different elements, the degrees of oxidation being indicated solely by the prefixes." *

The ending *ous* indicates a lower degree of oxidation than the ending *ic*.

15. *Acids.* — Very many of the non-metallic oxides combine with *water*, forming compounds which turn blue litmus-paper red. Such oxides are called *acid oxides*, or *anhydrides*, and their compounds with water are called *acids*. Thus $SO_2 + H_2O = H_2SO_3$. SO_2 is *sulphurous anhydride*, often improperly called *sulphurous acid*, a name which properly belongs to H_2SO_3. So $H_2O + SO_3$ (*sulphuric anhydride*) $= H_2SO_4$ (*sulphuric acid*).

Here, too, as in the case of oxides, the ending *ous* denotes a lower degree of oxidation than the ending *ic*.

* First Principles of Chemical Philosophy. By Josiah P. Cooke, Jr., 1868, p. 69.

When more than two acids are formed by the same elements, the prefixes *hypo* (less) and *hyper* or *per* (more) are used. Thus, hydrogen, chlorine, and oxygen form four acids: $HClO$, *hypochlorous acid;* $HClO_2$, *chlorous acid;* $HClO_3$, *chloric acid; and* $HClO_4$, *hyperchloric acid*, or *perchloric acid.*

It will be noticed that these acids are named from the element with which the oxygen combines to form an acid oxide; *sulphuric* and *sulphurous* acid from *sulphur*, and so on. It will be seen also that in the symbols of the acids the *hydrogen* is placed *first*, and the *oxygen last.*

16. *Bases.*—Many of the metallic oxides combine with water to form compounds whose properties are the opposite of those of the acids. These are called *basic oxides*, and their compounds with water are called *bases.* The specific name of a base is made an adjective in *ic* or *ous* formed from the Latin name of the metal, followed by the term *hydrate.* Thus, $\frac{1}{2}(H_2O + K_2O) = HKO$. K_2O is *potassic oxide*, and HKO is *potassic hydrate.* So $H_2O + BaO$ (*baric oxide*) $= H_2BaO_2$ (*baric hydrate*).

The *potassic* and *sodic hydrates* (HKO and $HNaO$) and some others are called *alkalies.* They differ from the remaining hydrates in that they cannot be decomposed by heat.

17. *Neutrals.*—Those oxides which, like water, are neither acid nor basic, are called *neutral oxides*, or *neutrals.*

18. *Salts.*—A metal may take the place of the hydrogen in an acid, and the compound thus formed is called a *salt.* Thus, the metal *potassium* may take the place of the hydrogen in H_2SO_3 and H_2SO_4, forming the *salts* K_2SO_3 and K_2SO_4. A salt is named from the acid, the ending *ic* being changed to *ate*, and the ending *ous* to *ite*, with the name of the metal prefixed with the ending *ic* or *ous.* Thus, K_2SO_3 is *potassic sulphite*, and K_2SO_4,

potassic sulphate. $FeSO_4$ is *ferrous sulphate*, and $Fe_2 3SO_4$, *ferric sulphate.*

19. *Binary and Ternary Compounds.* — Compounds like the oxides, which contain *two* elements, are called *binary* compounds; and compounds like the acids mentioned above (15) which contain *three* elements, are called *ternary* compounds.

HYDROGEN.

20. *Preparation of Hydrogen.* — The most convenient way of preparing hydrogen is by the action of zinc on dilute sulphuric acid. The apparatus required is shown in Figure 7.

Fig. 7.

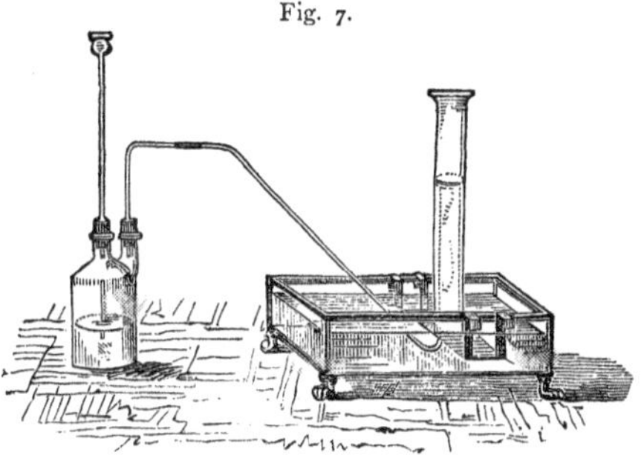

The reaction between the zinc and the sulphuric acid is shown in the following equation: —

$$H_2SO_4 + Zn = Zn\ SO_4 + H_2.$$

The zinc changes places with the hydrogen, forming the salt known as *zincic sulphate*, or *white vitriol*, while the hydrogen is set free.

21. *Properties of Hydrogen.* — If soap-bubbles are blown with a mixture of two measures of hydrogen to one of oxygen, they will explode violently when touched with a lighted taper; showing the intense energy with

which hydrogen and oxygen combine at a high temperature. If a mixture of hydrogen and atmospheric air be used, a similar explosion will take place. Hence, in preparing hydrogen for experiments, great care must be taken to expel all air from the apparatus before any of the gas is collected.

If a small rubber balloon be filled with hydrogen, it rises readily in the air, showing that the gas is lighter than air. It is, in fact, the lightest substance known, being 14.5 times lighter than air, and 16 times lighter than oxygen. It was formerly used in filling balloons; but coal-gas, which, though not so light, is much cheaper, is now commonly employed instead.

The lightness of hydrogen and the energy with which it combines with oxygen, are its two most important characteristics.

22. *The Oxy-hydrogen Blowpipe.* — The energy with which hydrogen and oxygen combine develops intense heat. This may be shown by means of the *oxy-hydrogen blowpipe*, represented in Figure 8. The hydrogen is forced through the tube *H*, and the oxygen through the tube *O*; and it will be seen that the two can mix only at the mouth of the jet where they are burned, so that there is no danger of an explosion. The gases may be kept in a *gasometer* (Figure 9), or, more conveniently, in gas-bags of india-rubber.

Fig. 8.

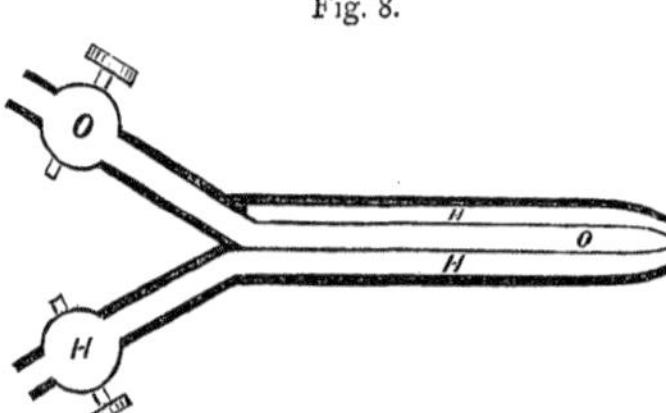

Hold a copper or iron wire in the flame of this blowpipe, and it burns as readily as a pine shaving held in the flame of a lamp. A steel watch-spring burns with brilliant scintillations. If a bit of zinc be placed on a piece of charcoal hollowed out for the purpose, and the flame of the blowpipe be directed upon it, the metal quickly melts and burns. Antimony, bismuth, and many other metals will burn in the same way, each with a characteristic light. Cast-iron burns with a shower of bright sparks. Platinum, which is one of the most infusible of substances, is readily melted, and even converted into vapor.

Fig. 9.

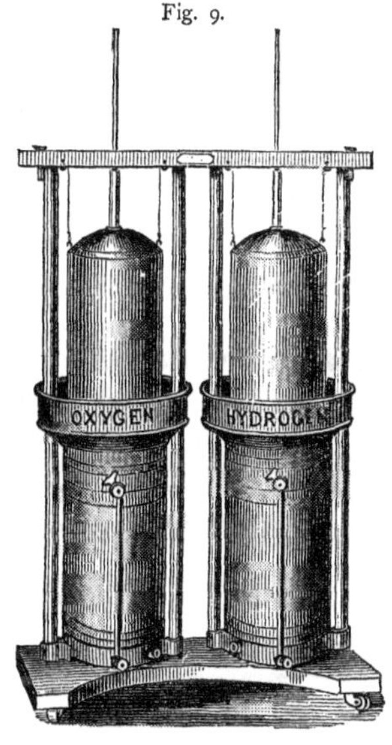

A small cylinder of lime placed in the blowpipe flame becomes white-hot, and glows with most intense brilliancy, forming what is known as the *lime light* or *Drummond light*. When the rays of this light have been gathered and reflected by a mirror, it has been seen at a distance of one hundred miles in full daylight.

WATER.

23. *The Properties of Water.*—The most important compound of hydrogen is *water*, which, as we have already learned (2), contains hydrogen and oxygen.

At ordinary temperatures, water is a clear, colorless transparent liquid, without taste or smell. It freezes at 32°, and under the ordinary pressure of the atmosphere it boils at about 212°.

It is the most universal *solvent* known. There are but few substances which are not dissolved by it. Even those which we call insoluble, generally differ from the rest only in degree. Spring water dissolves lime-rocks, and almost all other mineral substances. The beautiful crystals found in the rocks are almost invariably deposited from a solution of the mineral in water, though in many cases it may have taken thousands of years to form them.

Water dissolves almost all *gases* as well as solids. It is on the gases dissolved in the water that all aquatic plants and animals live.

24. *Various kinds of Natural Waters.* — Owing to its solvent powers, water is never found in nature in a state of purity. *Rain water*, collected after long-continued wet weather, comes nearest to it; but even this always contains small quantities of atmospheric air, and the gases floating in the air.

Spring water always contains more or less of saline matter, which it has dissolved from the soil. The salts most frequently found in it are common salt, calcic carbonate and sulphate (carbonate and sulphate of lime), and magnesic carbonate and sulphate (carbonate and sulphate of magnesia). Most spring water contains carbonic acid, to which it owes its sprightly taste.

Mineral waters are waters which contain an unusually large proportion of any of the salts just named, or of other and rarer substances. In many cases they have medicinal qualities, which vary with the salts held in solution. They sometimes contain salts of iron, and are then called *chalybeate* springs. In other instances, carbonic acid is so abundant as to give them an *effervescent* character. Less frequently *sulphur* is the chief ingredient, giving the water a nauseous taste and smell.

Water is familiarly called *hard* or *soft*, according to its action on soap. Hard waters contain salts of calcium or magnesium, which cause the soap to *curdle*, that is, to become insoluble. Soft waters do not contain these salts, and dissolve the soap without difficulty. Many hard waters become soft by boiling, the salts held in solution being deposited, as a *fur* or crust, on the inside of the boiler.

Sea water contains a large amount of common salt and of magnesic chloride (chloride of magnesium), to which it owes its saline, bitter taste. Smaller quantities of many other salts are found in it. All this matter has been washed out of the soil by the rivers which flow into the sea. It remains in the sea, since the water which evaporates from the surface is almost perfectly pure; but the surplus is continually taken up by marine plants and animals, so that the sea becomes no salter.

25. *Water in Plants and Animals.*—Water constitutes the greater part of all plants and animals. The human body is four-fifths water. In many of the lower animals the proportion is much greater. From a sun-fish weighing 30 pounds, only 240 grains of solid matter were obtained; so that water makes up about .999 of the weight of such animals. The vegetable substances which we use for food contain almost as large a percentage of water. In potatoes, the fraction is .75; in apples, .80; in turnips, .90; in watermelons, .94; and in cucumbers, .97.

26. *Water of Crystallization.*—Many salts, in crystallizing from their solutions, unite with a definite quantity of water, which is then called *water of crystallization.* If the salt be heated, the water is driven off, and the crystals fall to pieces; but the chemical properties of the substance are not altered. Many salts part with this water by mere exposure to the air, and *effloresce;* that is, crumble into white powder. Sodic carbonate (carbonate of soda) is an efflorescent salt. Other salts, on the contrary, like potassic carbonate (carbonate of potash), absorb water from the air, and become moist, or even dissolve; in which case they are said to *deliquesce.*

27. *Uses of Water.*—"On the uses of water it is almost needless to enlarge, for they are universally felt and appreciated. In each of its three physical conditions, the blessings which it confers on man are inestimable. As ice, it furnishes in northern lands, for months together, a solid bridge of communication between distant places: in the liquid condition, it is absolutely necessary to the existence of vegetable and animal life; in this shape, too, it furnishes to man a continual source of power in the flow of streams and rivers; it supplies one of the most convenient channels of communication between places widely separated; and further, it is the storehouse of countless myriads of creatures fitted for use as food: in the state of vapor, as applied in the steam-engine, it has furnished a power which has, in late years, done more than any other physical agent to advance civilization, to economize time, and to ameliorate the social condition of men. In each and all of these points, if rightly considered, we must perceive the entire adaptation of this compound to the ends which it was designed by the Creator to fill.

"Glancing at the physical condition of our planet, we cannot fail to be impressed with the important effects produced by the movements of water at periods anterior to the existence of man, as well as in more recent times. To such causes must we refer the formation of sedimentary rocks, and their arrangement in successive strata upon the surface of the earth: even now, observation shows that denudation is proceeding at some points, elevation and filling-up at others; whilst the accumulation of drift, and a variety of other extensive geological changes, must be traced to the same ever-acting and widely-operating agency.

"It may further be observed, that there is no form of matter which contributes so largely as water to the beauty and variety of the globe which we inhabit. In its solid state, we are familiar with it, in the form of blocks of ice, of sleet and hail, of hoar-frost fringing every shrub and blade of grass, or of snow protecting the tender plant, as with a fleecy mantle, from the piercing frosts of winter. The rare but splendid spectacles of mock suns, or *parhelia*, are due to the refractive power of floating spiculæ of ice upon the sun's rays. In its liquid condition, as rain or dew, it bathes the soil; and the personal experience of all will testify to the charm which the waterfall, the rivulet, the stream, or the lake, adds to the beauty of the landscape; whilst few can behold unmoved the unbounded expanse of ocean, which, whether motionless, or heaving with the gently undulating tide, or when lashed into fury by the storm which sweeps over its surface, seems to remind man of his own insignificance, and of the power of Him who alone can lift up or quell its roaring waves. In vapor, how much variety is added to the view by the mist or the cloud, which, by their ever-changing shadows, diversify, at every movement, the landscape over which they are flitting; whilst the gorgeous hues of the clouds around the setting sun, and the glowing tints of the rainbow, are due to the refractive action of water and watery vapor upon the solar rays." *

NITROGEN.

28. *Preparation of Nitrogen.* — Pure nitrogen may be readily prepared by passing a stream of chlorine gas through strong aqua ammonia. It is well to pass the nitrogen through a *wash-bottle*, to remove the white fumes, which, as we have seen (3), are a compound of hydrogen and nitrogen. The apparatus used is shown in

* Miller's Elements of Chemistry, Part II. pp. 35, 36.

Figure 10. The chlorine is generated in the flask at the left; the ammonia is put in the next bottle; and the nitro-

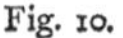

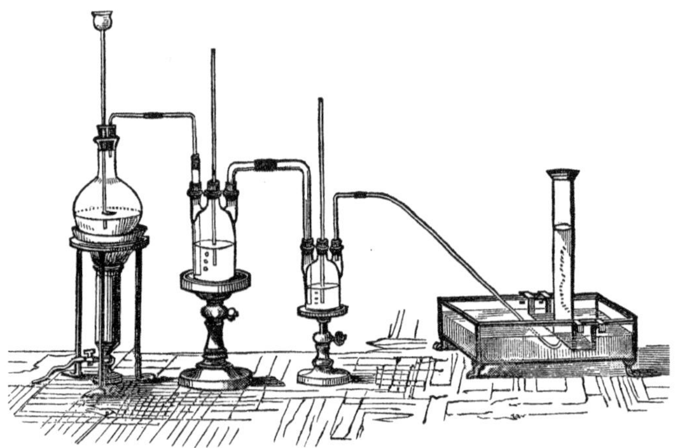

gen is washed in the third bottle, by being made to bubble up through water.

29. *Properties of Nitrogen.* — Nitrogen is a colorless gas, and is a little lighter than air. Its chief characteristic is its *inertness.* We have already seen (3) that it neither burns nor allows a taper to burn in it.

30. *Nitric Oxide.* — Put some bits of copper into a bottle (Figure 11), and add *aquafortis* or *nitric acid.* A

Fig. 11.

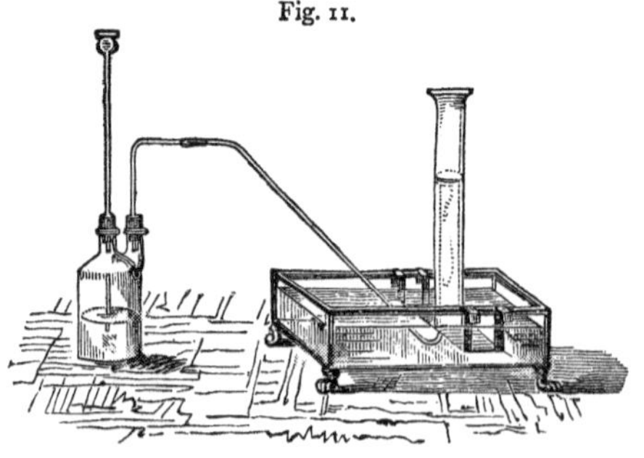

reddish gas fills the upper part of the bottle, and passes off through the tube. When collected over water, it is colorless. The red color was owing to impurities, which have been absorbed by the water.

Invert a jar of this gas over a piece of well-lighted phosphorus, and the latter burns almost as brilliantly as in oxygen, filling the jar with white fumes of *phosphoric anhydride* (*phosphoric acid.*) The experiment proves the presence of oxygen in the gas in which the phosphorus was burned.

Pass a mixture of equal parts of this same gas and hydrogen through a tube containing some heated *platinized asbestos*, and hold a piece of moistened red litmus-paper at the mouth of the tube. The paper is turned blue, showing that *ammonia* has been produced; for ammonia is the only gas that turns red litmus-paper blue. We have seen (3) that ammonia is a compound of hydrogen and nitrogen; therefore there must have been nitrogen in the gas mixed with the hydrogen.

The gas is, in fact, a compound of nitrogen and oxygen, and is called *nitric oxide* (*deutoxide* or *binoxide of nitrogen*), and its symbol is NO.

31. *Catalysis.* — In the above experiment the platinized asbestos remains wholly unchanged. By its mere *presence*, it has made the hydrogen unite with the elements of the gas mixed with it. This remarkable action by mere presence is called *catalysis.* It occurs in many other cases, but is very imperfectly understood.

32. *Nascent State.* — Nitrogen, as we have seen, is a very *inert* substance. Indeed, it can be made to combine *directly* with scarcely any element. In the above experiment, the hydrogen first seizes upon the oxygen, forming water, and sets the nitrogen free. At the moment when it is thus set free it has an increased activity, and readily unites with the hydrogen to form ammonia.

It is found generally true that elements are unusually active when just liberated from compounds. They are then said to be in the *nascent state.*

33. *Nitrous Anhydride.* — If a jar of oxygen be inverted over a jar of *nitric oxide*, the gases, on mixing, become of a bright cherry-red color. The compound formed differs from nitric oxide only in containing more oxygen. It is called *nitrous anhydride* (*nitrous acid*), and its formula is N_2O_3.

34. *Nitric Peroxide and Nitric Anhydride.* — It is possible to form two higher oxides of nitrogen, NO_2, *nitric peroxide*, and N_2O_5, *nitric anhydride.*

The former, at a temperature below 70°, is an orange-colored liquid; and the latter is a white solid, which eagerly unites with water, forming the well-known *aqua-fortis*, or *nitric acid*, HNO_3. The reaction between *nitric anhydride* and water is shown by the following equation: —

$$\tfrac{1}{2}(H_2O + N_2O_5) = HNO_3.$$

35. *Nitric Acid.* — Nitric acid is a strongly fuming liquid, colorless when pure, but usually yellowish from the presence of some of the lower oxides of nitrogen. It is one of the most corrosive substances known, and rapidly destroys all animal substances. When diluted, it stains the skin, wool, feathers, and the like, a bright yellow color, and is often used as a permanent yellow dye for silk and woollen goods. It acts violently upon tin and iron filings, especially when they are moistened with water. Indeed it attacks all the ordinary metals, except gold and platinum. Its action upon them, however, is more energetic when it is somewhat dilute.

The action of nitric acid upon the metals is owing to the readiness with which it parts with some of its oxygen. It first oxidizes them, and then acts upon these basic oxides (16), to form salts (17). During this process, the lower oxides of nitrogen are always given off, as is indicated by the appearance of red fumes.

Nitric acid may be made on a small scale by heating potassic nitrate (saltpetre), in a retort with sulphuric acid (oil of vitriol). The nitric acid distils over, and is collected in a receiver. The re-action is as follows: —

$$2KNO_3 + H_2SO_4 = K_2SO_4 + 2HNO_3.$$

The hydrogen and the potassium change places, giving rise to potassic sulphate (sulphate of potash) and nitric acid.

On the large scale, iron retorts, coated with fire-clay on the inside of the upper part, where they are exposed to the acid vapors, are employed for the distillation, and sodic nitrate (soda saltpetre) is substituted for potassic nitrate, as it is a cheaper salt, and likewise yields 9 per cent more nitric acid than potassic nitrate.

36. *Nitrous Oxide.* — A fifth oxide of nitrogen is obtained by gently heating ammonic nitrate (nitrate of ammonia). It is called *nitrous oxide*, and its symbol is N_2O. The reaction is shown by the following equation: —

$$(H_4N)NO_3 = 2H_2O + N_2O.$$

The ammonic nitrate, $(H_4N)NO_3$, breaks up into water and nitrous oxide.

N_2O is a colorless gas, and readily gives up its oxygen. Phosphorus burns in it as brilliantly as in pure oxygen. A splint of wood with a spark on the end of it bursts into flame when plunged into the gas. When inhaled, it has a peculiarly exhilarating effect, and hence is called *laughing gas.* It is used by dentists as an *anæsthetic*, since it makes the patient for the time insensible to pain. Great care must be taken that the gas when inhaled is ***perfectly pure.***

37. *Ammonia.*—We have seen (3) that ammonia is a compound of hydrogen and nitrogen. It is a colorless gas, which is greedily absorbed by water, forming the ordinary *aqua ammonia.* The eagerness with which it is taken up by water may be shown by filling a jar with the gas and opening it under water. The gas is instantly absorbed, and the water rushes in and fills the bottle. At a temperature of 32°, water will absorb more than one thousand times its bulk of ammonia.

Ammonia takes its name from the fact that it was first obtained from a salt found in Libya, near the temple of Jupiter Ammon. It is now prepared from the same salt, *sal-ammoniac* or *ammonic chloride*, which is made in large quantities from the waste liquor of gas-works. The sal-ammoniac is powdered and mixed with lime, and the mixture is heated in a flask arranged as shown in Figure 12. The gas is washed by passing it through a wash-bottle, and is then conducted into water.

Fig. 12.

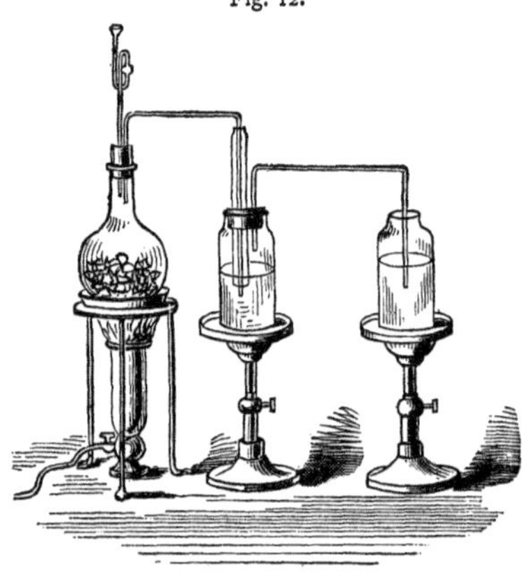

38. *Ammonium.*—The symbol of *ammonic chloride* is $(H_4N)Cl$; that of *ammonic nitrate* is $(H_4N)NO_3$. If we compare these with the symbols of the potassic chloride and nitrate, KCl and KNO_3, we shall see that the group of atoms H_4N plays the same part in the compounds as the *metal* potassium; and this is the case in a large class of salts. Hence this group of atoms has come to be considered as a *metal*, and has received the

name of *ammonium*. It has not, however, been obtained in a separate form, since it at once breaks up into ammonia, H_3N, and hydrogen.

39. *Cyanogen.* — When carbon and nitrogen are heated together in the presence of potash, a remarkable compound called *potassic cyanide* (*cyanide of potassium*), KCN, is formed. From this a large number of substances can be prepared, all of which contain the group of atoms CN. To this group the name *cyanogen* (blue-maker) is given, from its forming a number of blue compounds. It is remarkable as entering into composition exactly like a non-metallic element. It can be obtained in a free state as a colorless gas by heating mercuric cyanide (cyanide of mercury.) It forms with hydrogen an acid, HCN, analogous to muriatic acid, HCl. This acid is called *hydrocyanic* or *prussic acid*.

The most important of the blue compounds of cyanogen is *Prussian blue*, which may be formed by mixing a solution of either the *red* or the *yellow prussiate of potash* with a solution of *green vitriol*.

Groups of atoms, like H_4N and CN, which play the part of elements, are called *radicals*.

40. *The Characteristics of Nitrogen.* — The most marked features of nitrogen are its *inertness* (29); the *indirect way in which it enters into combination;* and the *great variety*, the *remarkable nature*, and the *instability of its compounds*.

The *great variety* of its compounds is shown in the five which it forms with oxygen: —

Nitrous oxide,	N_2O.
Nitric oxide,	NO.
Nitrous anhydride,	N_2O_3.
Nitric peroxide,	NO_2.
Nitric anhydride,	N_2O_5.

These compounds, as we have seen, are widely unlike in their properties; but in *composition* they differ only in the amount of oxygen which they contain.

The *remarkable nature* of the compounds is illustrated by *ammonium* and *cyanogen*, one of which acts like a *metallic* and the other as a *non-metallic element.*

The *instability* of the compounds is seen in the readiness with which the oxides give up their oxygen. *Nitric iodide* (*iodide of nitrogen*), which may be prepared by the action of ammonia on iodine, is so unstable that it explodes even when touched with a feather. *Nitric chloride* (*chloride of nitrogen*) is even more explosive, and is one of the most dangerous compounds known.

CARBON.

41. *Properties of Carbon.* — Carbon differs from the elements already described in being a solid when in a free state. It exists in three forms, which in outward appearance have nothing in common, but which are identical in their chemical relations. These three *allotropic* forms are *diamond*, *graphite* or *plumbago*, and *charcoal*. These differ in hardness, color, specific gravity, and the like; but when burnt in oxygen, the same weight of each produces the same weight of carbonic acid.

The *diamond* was first found to be pure carbon by Lavoisier in 1775–76. It occurs crystallized in certain rocks and gravels in India, Borneo, and Brazil. It is the hardest of all known bodies; and when cut, has a brilliant lustre and a high refractive power.

Graphite, *plumbago*, or *black lead*, is the substance used in the manufacture of the so-called "lead pencils." Although very brittle, its particles are very hard, so that a saw or other instrument used in cutting it is soon dulled. Mixed with oil it is extensively used to diminish the friction of machinery. It is the basis of most kinds of *stove-polish*, and of certain paints used to protect iron-work. It is also made into crucibles, which, on account of their infusibility, are much prized by the chemist. Carbon, in all its forms, is the most infusible substance known. Its *infusibility* and its *allotropic states* are its chief characteristics.

42. *Carbonic Anhydride* (*Carbonic Acid*).—The most important compound of carbon is CO_2, *carbonic anhydride*, commonly called *carbonic acid.* This gas is readily prepared by the action of dilute muriatic acid on

Fig. 13.

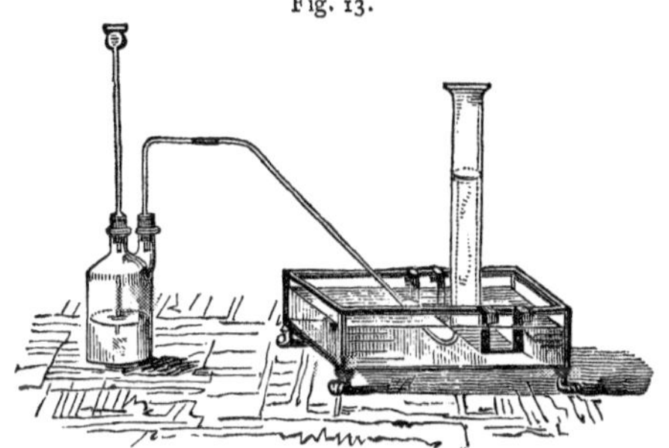

chalk or marble (Figure 13). The reaction is as follows:—

$$CaCO_3 + 2HCl = CaCl_2 + H_2O + CO_2.$$

$CaCO_3$, *calcic carbonate* (chalk, or marble), is broken up by the action of the acid. The calcium combines with the chlorine of the acid, forming *calcic chloride* (*chloride of calcium*), $CaCl_2$; one atom of the oxygen combines with the hydrogen of the acid, forming *water;* and CO_2 is set free.

We have already seen (11) that lime-water readily shows the presence of carbonic acid, by becoming milky white. The lime-water contains lime, CaO, which combines with the carbonic acid, CO_2, to form $CaCO_3$ again.

If a lighted taper be plunged into a jar of carbonic acid, it goes out. We have seen that the same thing takes place in nitrogen; but the two gases can be distinguished by the lime-water test.

Carbonic acid is about 1.5 times as heavy as air, so that, with an ordinary dipper, it can be dipped out from one jar and poured into another, like water. It is poisonous when breathed, even though mixed with a large quantity of air.

43. *Carbonic Oxide.* — Carbonic oxide, CO, is a gas formed when carbon burns with a limited supply of oxygen. It is often produced in an ordinary coal-fire, and is seen to burn with a pale blue flame. It is far more poisonous than carbonic acid.

CO may be obtained pure from several compounds of carbon. Thus, if crystallized oxalic acid, $C_2H_2O_4$, be heated with strong sulphuric acid, the latter withdraws from it one molecule of water, H_2O, leaving C_2O_3, which at once splits up into CO and CO_2. The CO_2 may be removed by passing the mixed gases through a solution of caustic soda.

SULPHUR.

44. *Sources of Sulphur.* — Sulphur occurs in nature both free and combined: it is found free in certain volcanic countries, especially in Sicily and Iceland; it exists in combination with many metals, forming *sulphides.* These sulphides are the ores from which several of the metals are commonly obtained. Thus PbS, plumbic sulphide (galena), ZnS, zincic sulphide (zinc blende), and CuS, cupric sulphide (sulphide of copper), are the most productive ores of those metals.

In order to obtain pure sulphur, the mineral, containing

the crude substance mixed with earthy impurities, is heated in earthen pots (Figure 13); the sulphur distils over in the form of vapor, which is condensed in similar pots placed outside the furnace. The sulphur thus obtained is refined or purified by a second distillation.

If the vapor of sulphur is quickly cooled below its melting-point, it solidifies in a crystalline powder, called *flowers of sulphur;* just as watery vapor, when cooled below the freezing point of water, becomes snow. If

Fig. 14.

sulphur is gently heated, it melts, and may be cast into sticks, and is then known as *brimstone*, or *roll sulphur.*

45. *Properties of Sulphur.* — Sulphur is a yellow solid, and exists in three allotropic forms (14). The first is that in which it crystallizes in nature. The other two are obtained by melting sulphur. If melted sulphur be allowed to cool slowly, it crystallizes in long, transparent needle-shaped crystals, which are quite different in form from the natural crystals, and have a specific

gravity of 1.98, while the natural crystals have a specific gravity of 2.07. These transparent crystals become opaque after an exposure of a few days to the air, owing to their breaking up into several crystals of the natural and permanent form. The third allotropic form of sulphur is obtained by pouring melted sulphur at 450° into cold water. It then forms a soft, tenacious mass, resembling india-rubber, and has a specific gravity of 1.96. This form of sulphur is not permanent. In a few hours it returns to its ordinary brittle state. These peculiar modifications are also apparent when sulphur is heated. It begins to melt at 239°, and forms an amber-colored limpid liquid; as the temperature rises, it begins to thicken, and to become dark-colored, so that at about 450° it can scarcely be poured from the vessel; heated still higher, it again becomes fluid, and remains as a dark thin liquid, till the temperature rises to 836°, when it begins to boil, and gives off a red vapor.

Sulphur is an inflammable substance, and when heated in the air or oxygen burns with a bluish flame, and with the suffocating odor so well known in the burning of a common lucifer match. It is insoluble in water and in most liquids, but dissolves in carbonic bisulphide (50).

46. *Sulphurous Anhydride* (*Sulphurous Acid*). — This compound, SO_2, as we have seen (11), is formed when sulphur is burned in oxygen, or in the air, which is a mixture of oxygen and other gases. The suffocating odor which we perceive when a common friction match is lighted is caused by SO_2.

The most important property of SO_2 is its power of bleaching silk, woollen, straw, etc. It is largely used for this purpose in the arts.

This property may be illustrated by holding a red rose or other flower in a jar of SO_2. The flower quickly loses its color.

47. *Sulphuric Acid.* — Sulphuric acid, or *oil of vitriol*, is the most important and most powerful of all the acids. It has a very strong attraction for water, and when the two are mixed, intense heat is developed. Hence caution is necessary in mixing them, especially in thick glass vessels. The acid should always be poured into the water, not the water into the acid. It rapidly chars all organic matter, and is a valuable drying agent.

Sulphuric acid was first prepared by distilling *green vitriol*, or *ferrous sulphate*, whence its name, *oil of vitriol.* In this form it is sometimes known as *Nordhausen* acid, from the name of the city in Saxony where it was first manufactured. It is a fuming liquid, denser than the ordinary acid, and is a mixture of this acid with the *sulphuric anhydride.*

This method has long been superseded by a more convenient one, which depends upon the fact that, though SO_2 does not combine with free oxygen and water to form sulphuric acid, it is capable of taking up the oxygen when the latter is united with nitrogen in the form of *nitrous anhydride*, N_2O_3. Thus: —

$$SO_2 + H_2O + N_2O_3 = H_2SO_4 + 2NO.$$

Sulphurous anhydride, water, and nitrous anhydride, yield sulphuric acid and nitric oxide.

The NO formed in this decomposition takes up another atom of oxygen from the air, becoming N_2O_3, and this is again able to convert a second molecule of SO_2, with water, H_2O, into sulphuric acid, H_2SO_4; being a second time reduced to NO, and ready again to take up another atom of oxygen from the air. Hence it is clear that the NO acts simply as a carrier of oxygen between the air and the SO_2; and a very small quantity of it is therefore able to convert a very large quantity of sulphurous anhydride, water, and nitrous anhydride, into sulphuric acid.

This process is conducted on a large scale in chambers made of sheet-lead, which are often of a capacity of 50,000 or 100,000 cubic feet. The sulphurous anhydride is obtained either by burning sulphur in a current of air, or by roasting a mineral called *iron pyrites* (ferric bisulphide, FeS_2), and is led, together with air, into the chamber. The N_2O_3 is got from sodic nitrate (soda saltpetre), which is decomposed either by the heat of the burning sulphur, or in a separate furnace. Jets of steam are also blown into the chamber, and a thorough draft is maintained by a high chimney. The sulphuric acid, as it forms, falls to the floor of the chamber, whence it is continually drawn off. It is then heated, first in open leaden pans, and then in vessels of glass or platinum (as lead is attacked by the strong acid), until the excess of water is driven off, and a sufficiently concentrated acid is obtained.

So extensively is this acid used, that the quantity manufactured in South Lancashire (England) alone exceeds 3,000 tons a week.

This method of making sulphuric acid may be illustrated, on a small scale, by dipping a shaving or cloth in nitric acid, and then holding it in a jar of sulphurous anhydride, with a little water at the bottom of the jar. Red fumes at once appear, which show that the nitric acid has lost some of its oxygen, having given it up to the SO_2, which thus becomes SO_3. If the experiment be repeated a number of times, the water at the bottom of the jar will be found to contain sulphuric acid.

48. *Sulphuric Anhydride.* — If fuming Nordhausen acid be gently heated in a glass retort, which is connected with a receiver, kept cool by ice, white fumes will pass over and solidify into a white silky fibrous mass. This solid is *sulphuric anhydride.* It has no acid properties, and can be moulded in the fingers with-

out charring the skin. When thrown on water, the heat emitted is so intense that it hisses like red-hot iron.

49. *Hydric Sulphide, Sulphuretted Hydrogen, or Hydrosulphuric Acid.* — Sulphur combines with hydrogen, forming *hydric sulphide*, H_2S, a compound analogous in composition to water. It is a gas of a very offensive odor, and very poisonous. It reddens litmus-paper, and is therefore an acid. It is best prepared by the action of dilute sulphuric acid on ferrous sulphide, FeS. Thus: —

$$FeS + H_2SO_4 = H_2S + FeSO_4.$$

The hydrogen and the iron change places, forming hydric sulphide and ferrous sulphate (sulphate of iron, or green vitriol).

This gas is an invaluable agent in chemical analysis.

50. *Carbonic Bisulphide.* — If the vapor of sulphur be passed over red-hot charcoal, a volatile compound, CS_2, is formed. It is seen to be analogous to carbonic acid, CO_2. It is a very inflammable, colorless liquid, of an offensive odor, and is called *carbonic bisulphide* (or *bisulphide of carbon*). It is of great importance in the arts, since it dissolves gums, caoutchouc, sulphur, and phosphorus.

51. *The Oxygen Group.* — The two compounds, H_2S and CS_2, are analogous in composition to H_2O and CO_2, and it is true in general that for every *oxide* there is a corresponding *sulphide*. Thus we have K_2O, *potassic oxide* (*potassa*), and K_2S, *potassic sulphide* (*sulphide of potassium*); BaO, *baric oxide* (*baryta*), and BaS, *baric sulphide* (*sulphide of barium*); Al_2O_3, *aluminic oxide* (*alumina*), and Al_2S_3, *aluminic sulphide* (*sulphide of aluminium*); and so on.

Oxygen and sulphur, then, belong to the same group of elements, which may be called the *oxygen* group. To this group also belong the rare elements, *selenium* and *tellurium*.

It will be noticed that the atomic weight of sulphur, 32, is just double that of oxygen.

CHLORINE.

52. *Preparation of Chlorine.* — Chlorine, as we have already seen (1), is prepared by heating a mixture of muriatic acid and black oxide of manganese, MnO_2. The reaction is shown in the following equation: —

$$MnO_2 + 4HCl = MnCl_2 + 2H_2O + 2Cl.$$

The hydrogen of the HCl combines with the oxygen of the MnO_2, forming two molecules of water, $2H_2O$. Half the chlorine of the HCl unites with the manganese, to form *manganic chloride* (*chloride of manganese*), $MnCl_2$, and the other half is set free.

53. *Properties of Chlorine.* — Chlorine was discovered by Scheele in 1774. It does not occur free in nature, but is found combined with metals, forming *chlorides*, of which rock-salt is the most common. Chlorine is, as we have seen, a greenish-yellow gas of a most disagreeable and irritating odor. It is very poisonous when inhaled in any considerable quantity. It is a very heavy gas, being nearly 2.5 times as heavy as air. Chlorine has a very strong affinity for the metals. If these in a finely divided state are brought into contact with it, they take fire spontaneously, giving rise to metallic chlorides. It has so strong an affinity for phosphorus, that this element will take fire spontaneously when put into a jar of chlorine. But the most remarkable property of chlorine is its power of combining with hydrogen to form muriatic acid. When these gases are mixed in equal volumes, they combine with an explosion

on applying a lighted taper to them, or on exposing them to the sunlight. This property of chlorine has already been illustrated in the decomposition of water and ammonia (2, 3).

Another remarkable property of chlorine is its *bleaching* power, which depends on its ability to decompose water. We may enclose a piece of cotton cloth colored with a vegetable substance in a bottle of dry chlorine, and no change of color takes place, even after many weeks. If, however, a few drops of water are added, the cotton is at once bleached. The chlorine combines with the hydrogen of the water, and the oxygen, at the moment of its liberation, when it is in the *nascent state* (32), combines with the coloring matters, forming colorless compounds. Ordinary free oxygen has not this power, at least to any great extent.

54. *Chloride of Lime, or Bleaching Powder.*— Since chlorine destroys all organic colors (that is all colors derived from the vegetable and animal kingdoms) it is very extensively used in bleaching cotton and linen goods and paper. Cotton fabrics may now be rendered perfectly white in a few hours; while, by the old method of laying them on the grass in the sun, it required weeks, and even months, to effect it. It was necessary, moreover, to have meadow-land suitably situated for the bleaching; hence most of the cloth manufactured in England was carried to Holland to be bleached. Besides the diminished expense, the cotton stuffs bleached with chlorine suffer less injury in the hands of skilful workmen than those bleached in the sun.

The health of the workmen is greatly endangered by the use of chlorine in the gaseous state; but it has been found that the gas is readily absorbed by slaked lime, and is as readily given up again when the lime is treated with dilute acid.

When thus combined with lime, the chlorine can be easily transported. The compound is called *chloride of lime*, or *bleaching powder*.

This powder is made on a large scale by conducting chlorine into spacious chambers, on the floor of which slaked lime is spread to the depth of two inches.

In bleaching, the goods are first dipped into a solution of the bleaching powder, and then passed through dilute acid. The chlorine is thus set free in the fibres of the cloth, where it does its work of bleaching without injury to the workmen.

Chloride of lime is also largely used as a *disinfectant*. The chlorine acts upon organic *odors* in the same way as upon organic colors; that is, it *oxidizes* and destroys them.

55. *Hydric Chloride* (*Hydrochloric Acid*).—*Hydric chloride* or *hydrochloric acid* is a colorless gas, somewhat heavier than air, and is very soluble in water. This solution is the ordinary *muriatic acid* of the shops. It may be easily prepared, by placing in a capacious retort 3 parts of fused sodic chloride (chloride of sodium, or common salt) in fragments, and adding slowly, through a bent funnel, 3 parts of oil of vitriol. If pounded salt be used, the action of the acid is apt to be too rapid. The retort is connected with a series of bottles; in the first, a small quantity of water is placed, to detain any impurities which might be carried over mechanically with the gas; the second bottle may contain 4 parts of water, and should be immersed in a vessel of cold water, as the condensation of the gas is attended with a great disengagement of heat. On applying a gentle heat to the retort, the acid comes over and is condensed; an easily soluble sodic sulphate (sulphate of soda) remains in the retort. For manufacturing purposes, the decomposition is effected in iron cylinders, like those employed in the manufacture

of nitric acid, and only one half the quantity of sulphuric acid prescribed above is used: —

$$2NaCl + H_2SO_4 = 2HCl + Na_2SO_4.$$

Sodic sulphate, Na_2SO_4, remains in the cylinder, whilst the acid is condensed in a series of salt-glazed stone-ware jars.

Enormous quantities of muriatic acid are obtained as an incidental product in the manufacture of soda-ash, which will be described farther on. More than 1000 tons are made every week at the soda-ash works in South Lancashire (England) alone.

56. *Hydracids.* — Hydric chloride, HCl, and hydric sulphide, H_2S, contain each but two elements, and no oxygen; while nitric acid, HNO_3, and sulphuric acid, H_2SO_4, contain each three elements, one of which is oxygen. There are, then, two classes of acids: one containing oxygen, and called *oxyacids;* and the other, containing no oxygen, called *hydracids.*

57. *Aqua Regia.* — When muriatic acid is mixed with nitric acid, it forms the so-called *nitro-muriatic acid.* This acid was called *aqua regia* (royal water) by the alchemists, because it dissolves gold, the "king of metals." Both platinum and gold are insoluble in either acid separately; but when the two acids are mixed, they decompose each other. Chlorine is set free, and in its nascent state (32) acts upon the metals, and dissolves them.

58. *Chlorine Oxyacids.* — Chlorine has a very weak affinity for oxygen; but it forms four oxyacids: *hypochlorous acid*, HClO; *chlorous acid*, $HClO_2$; *chloric acid*, $HClO_3$; and *hyperchloric* or *perchloric acid*, $HClO_5$. Of these, the hypochlorous and the chloric form some important salts.

BROMINE.

59. *Properties of Bromine.*—This element, which closely resembles chlorine in its properties and compounds, was discovered by Balard, in 1826, in the salts obtained by the evaporation of sea-water. It does not occur free in nature, and is, like chlorine, found combined with sodium and magnesium as *bromides* in certain mineral springs.

Bromine is a dark, reddish-black, heavy liquid, being the only element, except mercury, that exists as a liquid at the ordinary temperature. Its specific gravity is 2.966, it freezes at $-9°$, and boils at 145°. It has a very strong, irritating smell, resembling that of chlorine, and when inhaled, acts as a violent poison. It is quite soluble in water, and this solution bleaches, but more feebly than a solution of chlorine. Some of its salts are much used in medicine and in photography.

IODINE.

60. *Properties of Iodine.*—Iodine also occurs in sea-water, combined with sodium and magnesium. It was discovered in 1812 by Courtois. At the ordinary temperature it is a dark gray solid, with a bright metallic lustre. It has a specific gravity of 4.95; it melts at 225°, and boils at about 350°. Its vapor is of a beautiful deep violet color, and has a faint chlorine-like smell. Iodine does not possess such active qualities as either chlorine or bromine. Its solution does not bleach vegetable colors, and it is set free from its compounds by both the above elements. Free iodine forms with starch a remarkable compound of a rich blue color (12). Iodine acts as a violent poison; but, given in small quantities, it is much used as a medicine. Some of its salts also are used in medicine and in photography.

FLUORINE.

61. *Properties of Fluorine.* — This element occurs combined with the metal calcium, as calcic fluoride (fluoride of calcium) or *fluor-spar*. It is remarkable as forming no compound with oxygen, and as being extremely difficult to prepare in a pure state. Many unsuccessful attempts have been made to obtain fluorine; but, by the action of dry iodine upon dry argentic fluoride (fluoride of silver), it appears to have been isolated; and it is found to be a colorless gas.

62. *Hydric Fluoride, or Hydrofluoric Acid.* — When calcic fluoride (fluor spar) is acted upon by sulphuric acid, a very corrosive gas is obtained, called *hydric fluoride* (*hydrofluoric acid*). The reaction is shown in the following equation: —

$$H_2SO_4 + CaF_2 = CaSO_4 + 2HF.$$

The calcium and the hydrogen change places, forming *calcic sulphate* (*sulphate of lime*), $CaSO_4$, and hydric fluoride.

This acid must be prepared in vessels of lead or platinum, since it rapidly attacks glass. It acts violently upon the skin, producing painful sores, and the fumes are very poisonous to the lungs. It is used for etching on glass. The glass is first covered with wax, through which the design is traced with a sharp point. The glass is then exposed to the gas, which corrodes it where the wax has been removed. The solution of the gas in water is also used for the same purpose.

63. *The Chlorine Group.* — The four elements just described form a well-marked group. They are especially remarkable for their gradation of properties. Thus chlorine is a gas, bromine a liquid, and iodine a solid, at the ordinary temperature. Chlorine may be easily con-

densed into a liquid, and the specific gravity of liquid chlorine is 1.33, of bromine 2.97, and of iodine 4.95. Liquid chlorine is transparent, bromine slightly so, and iodine opaque. The atomic weight and density of bromine are nearly the mean of those of chlorine and iodine, $\frac{35.5 + 127}{2} = 81.25$; and in its general chemical deportment bromine stands half-way between the other two elements.

These elements all form powerful acids with hydrogen, and have a strong affinity for the metals. They all have a very weak affinity for oxygen.

PHOSPHORUS.

64. *Properties of Phosphorus.* — This element does not occur free in nature, but is found combined with oxygen and calcium in large quantities in the bodies, especially the bones, of animals, and in the seeds of plants. Animals obtain their phosphorus from plants, plants from the soil, and the soil from the slow disintegration of the oldest rocks. Phosphorus was accidentally discovered by Brande, in 1669; but Scheele, in 1769, pointed out its existence in the bones, and carefully examined its properties. It is prepared from bones, and is used chiefly in the manufacture of lucifer matches.

Phosphorus is a yellowish, semi-transparent solid, resembling wax. Its specific gravity is 1.83. It melts at 111°, and boils at 550°. In the air it gives off white fumes, and emits a pale light in the dark, whence its name, which means *light-bearer*. While emitting this light, it appears to be undergoing a slow burning. At a temperature a little above its melting-point, phosphorus takes fire in the air, and burns with great energy. It may be set on fire by slight friction, by a blow, and even by the heat of the hand. Hence it should be handled with great care, and always be cut under water. Phosphorus has several well-marked allotropic states, the most important of which is that known as *red phosphorus.* This is much less inflammable than ordinary phosphorus, and is not poisonous.

65. *Compounds of Phosphorus.*—Phosphorus combines with oxygen, to form several acids, the most important of which are known as *hypophosphorous acid* and *phosphoric acid.* From the former are derived salts (*hypophosphites*), used in medicine; while the latter furnishes a large class of salts (*phosphates* and *superphosphates*), which are valuable to the farmer as fertilizers. Phosphorus unites with hydrogen, to form *hydric phosphide* (phosphuretted hydrogen), H_3P, a poisonous gas, with a very offensive odor. When this gas is prepared by the action of caustic potash on phosphorus, each bubble of the gas takes fire as it comes into contact with the air, forming beautiful rings of phosphoric anhydride (phosphoric acid), which expand as they rise. This spontaneous combustion is owing to the presence of a small quantity of a liquid compound of phosphorus and hydrogen, whose symbol is H_2P. Pure phosphuretted hydrogen does not take fire in this way.

ARSENIC.

66. *Properties of Arsenic.*—Arsenic closely resembles phosphorus and nitrogen in its chemical properties, and in those of its compounds, and is therefore to be counted as a non-metallic element. In physical features, such as specific gravity and lustre, it bears a greater resemblance to the metals, with which it was formerly classed. (See, in Appendix, *Chemical Philosophy*, § 19.)

Arsenic is sometimes found in a free state, but more often combined, chiefly with iron, nickel, cobalt, and sulphur. One of the compounds of this element with oxygen, As_2O_3, is the well-known poison, *white arsenic.*

67. *The Nitrogen Group.* — Nitrogen, phosphorus, and arsenic form a well-marked group. They all form similar compounds with oxygen and with hydrogen. Thus we have H_3N (ammonia), H_3P (phosphuretted hydrogen), and H_3As (hydric arsenide); N_2O_3 (nitrous anhydride), N_2O_5 (nitric anhydride), P_2O_3 (phosphorous anhydride), P_2O_5 (phosphoric anhydride), and As_2O_3 (arsenious anhydride), As_2O_5 (arsenic anhydride).

BORON.

68. *Properties of Boron.* — Boron is found in nature combined with oxygen as *boracic acid*, and with oxygen and sodium as *borax*. In certain volcanic districts in Tuscany, jets of steam and gas are continually escaping from the earth. These steam-jets contain small quantities of boracic acid, which collects in artificial basins formed for the purpose at the mouth of the jet. By means of the heat of natural steam-jets, this solution of boracic acid is concentrated and crystallized. About 2,000 tons of crude acid are thus prepared and exported from Tuscany every year.

Boron is remarkable as combining directly with nitrogen at a high temperature.

SILICON.

69. *Properties of Silicon.* — Next to oxygen, silicon is the most abundant element known. It does not, however, occur in a free state, but always combined with oxygen (for which it has a very strong affinity) as *silica.*

Silica is found nearly pure in quartz, flint, sand, and in several minerals, and combined with metals in almost all known rocks. Its symbol is SiO_2.

Silicon can be obtained in three modifications, corresponding to the three states of carbon.

70. *The Carbon Group.* — Carbon and silicon, then, form a well-marked group, resembling each other closely in their allotropic states.

SUMMARY.

Chemistry divides substances into two classes: —

1st, *Elements*, or substances which cannot be decomposed by any known process. Of these only 65 are known.

2d, *Compounds*, or substances made up of elements.

The *force* which causes elements to combine, and holds them in combination, is called *Affinity*.

Affinity *changes the properties of the substances which it combines.*

Affinity *always causes substances to combine in fixed and definite quantities.* This law of combination is called the *Law of Definite Proportions.*

A given element combines with some elements in preference to others.

Matter is made up of *molecules;* and molecules are made up of *atoms.*

The compounds of nitrogen and oxygen show that the same elements may combine in more than one proportion, and that in such cases the proportions of the elements in the compounds are always *multiples* of the *atomic weights* of the elements. This law of combination is known as the *Law of Multiple Proportions.*

The symbol of an *element* is usually the first letter of its name, a second letter being added to distinguish names beginning with the same letter. The symbol of an element always stands for *one atom* of the element.

The symbol of a *compound* indicates its composition. It is formed by writing together the symbols of the elements of the compound, with a small figure after each symbol expressing the number of atoms of that element found in a molecule of the compound. The symbol of a compound always stands for *one molecule* of the compound.

The chemical changes, or *reactions*, which substances undergo, are indicated by *equations* made up of these symbols.

The compounds of oxygen with the elements are called *oxides*. These oxides are either *acid*, *neutral*, or *basic*. The oxides of the non-metallic elements are generally acid; those of the metallic elements are generally basic.

The acid oxides combined with water form *acids*.

The basic oxides combined with water form *bases*.

When a metal takes the place of the hydrogen of the acid, the compound is called a *salt*.

A *hydracid* is made up of hydrogen and a non-metallic element.

The names of *binary salts* take the ending *-ide;* those of *ternary salts*, the endings *-ate* or *-ite*, according as the name of the acid ends in *-ic* or *-ous*.

In general, the name of a binary compound takes the ending *-ide*.

Oxygen is the most abundant element, and is usually prepared from potassic chlorate. It has a wider range of affinities than any other element.

Ozone is an allotropic form of oxygen, and is more active than ordinary oxygen.

Hydrogen is the lightest element, and is usually obtained by the action of zinc on dilute sulphuric acid. Its most important compound is *water*.

Nitrogen is one of the most inert of elements. It is remarkable for the indirect way in which it enters into combination, and for the variety and the nature of its compounds.

Carbon is an infusible solid, and has three allotropic states.

Sulphur is an inflammable solid, forming *sulphides* which in their composition resemble oxides. It has several remarkable allotropic states.

Chlorine is most readily prepared by the action of black oxide of manganese on muriatic acid. It is distinguished by its color, its odor, and its affinity for hydrogen and the metals. Its bleaching and disinfecting power is due to its strong affinity for hydrogen.

Bromine, *iodine*, and *fluorine* are elements closely resembling chlorine in their chemical properties.

Phosphorus is a very inflammable substance, and is mainly used for making lucifer matches. It is obtained from bones.

Phosphorus and *arsenic* resemble nitrogen in the compounds which they form.

Silicon resembles carbon in its allotropic states.

Boron combines directly with nitrogen at a high temperature.

Nitric acid is remarkable for its *oxidizing* power.

Sulphuric acid is the strongest of acids, and is remarkable for its affinity for water.

Sulphurous acid is a bleaching agent.

Nitrous oxide is used as an anæsthetic.

Carbonic acid is a heavy gas, and is most readily prepared by the action of muriatic acid on marble.

The most important non-metallic *sulphides* are *hydric sulphide* and *carbonic bisulphide*.

The most important of the non-metallic *chlorides* is *muriatic acid*.

Cyanogen is a remarkable compound of nitrogen and carbon, and plays the part of a non-metallic element.

Ammonium is a remarkable compound of nitrogen and hydrogen, and plays the part of a metal.

Ammonia, although it is not an oxide, has very powerful basic properties. It is known as the *volatile alkali*, and is extensively used in the arts.

THE METALS.

GENERAL PROPERTIES OF THE METALS.

71. *Physical Properties of the Metals.* — The metals differ widely from one another, both in their physical and chemical properties. Those metals which are lightest have the greatest affinity for oxygen, whilst the heavier metals are oxidized with difficulty.

The following Table gives the specific gravities of the most important metals: —

Iridium	21.8	Iron	7.8
Platinum	21.5	Tin	7.3
Gold	19.3	Zinc	7.1
Mercury	13.596	Antimony *	6.7
Lead	11.3	Chromium	5.9
Silver	10.5	Aluminium	2.56
Bismuth *	9.8	Strontium	2.54
Copper	8.9	Magnesium	1.75
Nickel	8.8	Calcium	1.58
Cadmium	8.6	Sodium	0.972
Cobalt	8.5	Potassium	0.865
Manganese	8.0	Lithium	0.593

The melting-points of the metals differ even more widely than their densities, as will appear from the following Table, which is mainly compiled from Miller: —

* See in Appendix the chapter on *Chemical Philosophy*, §§ 19 and 20.

Mercury	—39°	Antimony	+737°
Cadmium	+242	Silver	1873
Tin	442	Copper	1996
Bismuth	507	Gold	2016
Lead	617	Steel	2350 to 2500
Zinc	733	Wrought Iron	2700 to 2850

Some metals can be easily converted into vapor, or volatilized; thus mercury boils at 662°, while potassium, sodium, magnesium, zinc, and cadmium can be distilled at a red heat. Even the more infusible metals, such as copper and gold, are not absolutely fixed, but give off small quantities of vapor when strongly heated.

72. *Metallic Ores.* — Only a few of the metals occur in a free state in nature; in general they are found combined with oxygen, sulphur, or some other non-metal. These metallic compounds are most variously distributed throughout the earth's crust; some are known to occur in only one or two localities, and even then only in minute quantity, while others are found widely distributed in enormous masses. Aluminium, iron, calcium, magnesium, and sodium form, when united with oxygen and silicon, the whole mass of granitic rocks composing our globe; but it is not from these sources that they can be obtained for the purposes of the arts. For this object we employ other combinations, found in smaller quantity, termed *metallic ores*, from which the metals can be more easily extracted.

The heavy metals and their ores are interspersed throughout the older rocks, in the form of *veins* or *lodes*, which are cracks or fissures of the rock filled up with the ore. Other ores, such as ironstone, are found in the more recent rocky formations, having been deposited in large masses, probably from aqueous solution.

THE MOST USEFUL METALS.

The most useful metals are *iron*, *lead*, *copper*, *tin*, and *zinc*.

73. *IRON.*—Iron is, of all metals, the most important to mankind. Its uses were long unknown to the human race, the age of iron implements being preceded by the ages of bronze and stone. Pure metallic iron exists only in very small quantity on the earth's surface, almost entirely occurring in meteoric stones.

The process of obtaining iron from its ores requires an amount of knowledge and skill which the early races of men did not possess. The iron of commerce exists in three different forms, exhibiting very different properties, and having different chemical constitutions: (1) *wrought iron;* (2) *cast-iron;* (3) *steel.*

The first is nearly pure iron; the second is a compound of iron with varying quantities of carbon and silicon; and the third a compound of iron with less carbon than is needed to form cast-iron.

Pure iron in the form of powder may be obtained by heating the oxide moderately in a current of hydrogen. The hydrogen combines with the oxygen to form water, leaving the iron free. It must, however, be kept in an atmosphere of hydrogen, as finely divided iron takes fire spontaneously in the air. Iron has a bright white color, and is remarkably tough, though soft. The pure metal crystallizes in cubes. Iron which has been hammered has, when broken, a granular structure; but it becomes fibrous when the iron is rolled into bars, and the more or less perfect form of the fibre determines to a great extent the toughness and the value of the metal. This fibrous texture of hammered bar-iron undergoes a change when exposed to long continued vibration, the iron returning to its original crystalline condition. Many accidents have occurred in the breaking of railway axles, owing to this change from the fibrous to the granular texture. Wrought iron melts at a very high temperature; but, as it becomes soft at a much lower point, it can be easily worked. When hot, it possesses the peculiar property of *welding;* that is, the power of uniting firmly when two clean surfaces of hot metal are hammered together.

Iron and certain of its compounds are strongly magnetic. The metal loses this power when red-hot, but regains it upon cooling. A solid mass of iron does not oxidize or tarnish in dry air, at the ordinary temperature; but if heated, it oxidizes in black scales. When more strongly heated in the air, or plunged into oxygen gas, it burns with the formation of the same black oxide (11). In pure water, iron does not lose its brilliancy; but, if a trace of carbonic acid is present, and access of air is permitted, the iron soon rusts.

74. *Manufacture of Iron.* — The oldest method of manufacturing wrought iron was to heat the ore in a blast-furnace with charcoal or coal, and to hammer out the spongy mass of iron thus obtained. This plan can only be economically followed on a small scale and with the purest ores, and has been superseded by a more complicated method, applicable to all kinds of iron-ore. The most important ore is an impure carbonate called *clay ironstone.* This is first *roasted* (that is, heated in contact with air) to drive off the carbonic acid, and reduce the iron to an oxide. It is next *smelted* in a *blast-furnace.* These furnaces (Figure 15) are usually about fifty feet high, and about fifteen feet in diameter in the widest part of the cavity *CD.* The lowest part, *F,* is called the *crucible,* or *hearth.* *I I* are the *tuyères,* or pipes through which air is forced by powerful bellows.

K, *K*, and *M* are arched galleries for the convenience of workmen employed about the furnace. When working regularly, the furnace is charged from a door at the end of the gallery near the top, first with coal, and then with a mixture of roasted ore and limestone broken into small pieces. As the fuel burns away and the materials gradually sink, fresh supplies of fuel and of ore are added; so that the furnace is kept filled with alternate layers of each.

Fig. 15.

The oxygen of the air from the bellows combines with the carbon of the fuel, forming carbonic oxide, CO, which rises through the porous mass, and, taking the oxygen from the ore, becomes converted into carbonic acid, CO_2. The iron, mixed with the earthy matter of the ore, settles down into the hottest part of the furnace, where both are melted. The iron, being the heavier, sinks to the bottom, where it is drawn off at intervals through a *tap-hole* in the floor *H*. The lighter earthy matter, or *slag*, floats on the surface of the iron, like oil on water, and flows off through an opening above the *tymp-stone L*. The limestone aids in liquefying the earthy matter, and unites with it to form the slag.

At this stage the iron is *cast-iron*.

The properties and appearance of cast-iron vary much

with the amount of carbon and silicon which it contains; for cast-iron is not a definite chemical compound of these elements with iron. The carbon is found in cast-iron, (1) as scales of graphite, giving rise to *mottled* cast-iron; and (2) in combination, forming *white* cast-iron.

In order to make *wrought iron* from cast-iron, the latter must undergo the processes of *refining* and *puddling*. These consist essentially in burning out the carbon and silicon, by exposing the heated metal to a current of air in a *reverberatory* furnace. Such a furnace is repre-

Fig. 16.

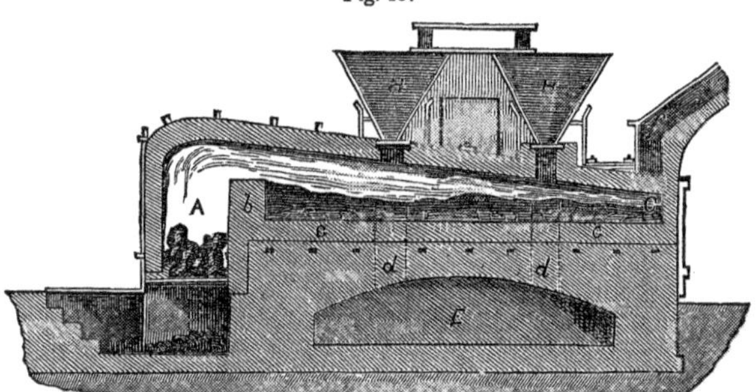

sented in Figure 16. The ore is put into the hoppers *H H*, from which it falls into the chamber *C*, where it is spread out on the *bed c c*. The fuel is burned on a hearth at *A*, separated from the ore by the bridge *b*. The heated gases rising from the burning fuel are *reverberated*, or reflected, by the arched roof of the furnace, and driven down upon the ore, and then pass off through the flue *f*. When the ore is sufficiently roasted, it is allowed to fall through openings, *d d*, into the chamber *E*. The ore is stirred from time to time, to expose fresh surfaces to the action of the air and the flame. The melted cast-iron becomes first covered with a coat of oxide, and gradually thickens, so as to allow of its being rolled into large lumps or balls. During this process, the whole of the carbon escapes as carbonic oxide, and the silicon becomes oxidized to silica, which unites with the oxide of iron, and forms a fusible slag. Any phosphorus or sulphur contained in the iron is also oxidized in this process. The ball is then hammered, to give the metal coherence, and to squeeze out the liquid slag, and the mass is afterwards rolled into bars or plates.

75. *Steel.* — Steel is formed when bars of wrought iron are heated to redness for some time in contact with charcoal. The bar is then found to have become fine-grained instead of fibrous; the substance is more malleable and more easily fusible than the original bar-iron, and is found to contain carbon varying in amount from one to two per cent. Steel has several important properties, especially the power of becoming very hard and elastic when quickly cooled, which fits it for the manufacture of edge-tools. These are, however, generally made of bar-steel, which has been previously fused and cast into ingots.

A new and very rapid mode of preparing cast-steel is that known as the *Bessemer process*. This process consists in burning out all the carbon and silicon in cast-iron, by passing a blast of atmospheric air through the molten metal, and then adding such a quantity of pure cast-iron to the wrought iron thus prepared as is necessary to give carbon enough to convert the whole mass into steel. The melted steel is then at once cast into ingots. In this way six tons of cast-iron can, at one operation, be converted into steel in twenty minutes. The Bessemer steel is now largely manufactured for railway axles and rails, for boiler-plates, and other purposes, for which it is much better fitted than wrought iron, so that this process bids fair to revolutionize the old iron industry. It has already been put into practice on a large scale in England, France, Belgium, Sweden, and India.

76. *LEAD.*—Lead does not occur free in nature. It is obtained from *galena*, or plumbic sulphide, PbS. The galena is roasted in a reverberatory furnace, with the addition of a small quantity of lime to form a fusible slag with any silicious mineral matter present in the ore. By the action of the air, a portion of the sulphide is oxidized to sulphate, while in another portion the sulphur burns off, and plumbic oxide is left behind. After a time, the air is excluded, and the heat of the furnace raised, and the remaining sulphur burns at the expense of the oxygen in the oxide already formed, leaving the lead free.

Lead is a bluish-white metal, and so soft that it may be scratched with the nail. It is quite ductile and malleable, but has little tenacity or elasticity. It melts at 617°, and at a higher temperature volatilizes, though not in quantity sufficient to be distilled.

The surface of the metal remains bright in dry air, but it soon becomes tarnished in moist air, owing to the formation of a film of oxide; and this oxidation proceeds rapidly in presence of a small quantity of weak acid, such as carbonic or acetic. In pure water freed from air, lead also preserves its lustre; but, if air be present, plumbic oxide is formed, and as this dissolves slowly in the water, a fresh portion of metal is exposed for oxidation. This solvent action of water upon lead is a matter of much importance, owing to the common use of lead water-pipes, and the peculiarly poisonous action of lead compounds upon the system when taken even in minute quantities for a length of time. The small quantity of certain salts contained in all spring and river waters exerts an important influence on the action of lead. Thus waters containing nitrates or chlorides are liable to contamination with lead, while those *hard* waters containing sulphates or carbonates may generally be brought into contact with lead without danger, as a thin deposit of sulphate or carbonate is formed, which preserves the metal from further action. If the water contains much free carbonic acid, it should not be allowed to come into contact with lead, as the carbonate dissolves in water containing this substance. The presence of lead in water may easily be demonstrated by acidulating the water, passing a current of hydric sulphide (sulphuretted hydrogen) through a deep column of the liquid, and noticing whether it becomes tinged of a brown color, from the formation of plumbic sulphide.

Lead pipes lined with tin have been recommended for water-pipes; but if the lining is not absolutely perfect, they are more dangerous than ordinary lead pipes. If from defective soldering, or cracks, or breaks in the lining, or from corrosive action, the water comes in contact with the lead, a galvanic action begins, and the lead is more rapidly oxidized than it would be if not joined with the tin.

There is the same danger in the use of the metallic double-lined ice-pitchers. The lining is often made of dissimilar metals, and the parts are joined by a solder containing lead. The thin film of silver is soon worn off upon the interior; galvanic action then promotes corrosive action, and the water becomes poisonous.

On the whole, the best way of protecting lead-pipes from oxidation, is by coating them with sulphide of lead, which is insoluble in water. This can be done by dissolving one pound of potassic sulphide (sulphide of potassium) in two gallons of water, and letting the solution remain in the pipe twelve hours, or until the whole inside is thoroughly blackened. This preventive process is not perfect, but it is very nearly so.

77. *COPPER.* — Copper is an important metal, largely used in the arts. It has been known from very

early times, as it occurs native in the metallic state, and is moreover easily reduced from its ores. Metallic copper is found in enormous quantity near Lake Superior, in North America, and other localities. The following ores are the most important: (1) a compound of copper, sulphur, and iron, known as *copper pyrites*, $Cu_2S + Fe_2S_3$; (2) the *cuprous sulphide*, Cu_2S; (3) the *carbonate* or *malachite;* and (4) the *red* or *cuprous oxide*, Cu_2O. Copper is obtained on a large scale from the carbonate or oxide, by reducing these ores, together with carbon and some silica, in a blast-furnace.

Metallic copper has a peculiar deep red color, which is best seen when a ray of light is several times reflected from a bright surface of the metal; it is very malleable, ductile, and tenacious; it melts at a red heat, and is slightly volatile at a white heat, giving a green tint to a flame of hydrogen gas which is passed over it; and it is one of the best conductors of heat and electricity. Copper does not oxidize, either in pure dry or moist air, at ordinary temperatures, but if heated to redness in the air, it is soon converted into cupric oxide.

78. *Alloys of Copper.*—Copper combines with several of the metals to form what are called *alloys.*

Brass is an alloy containing about two-thirds of copper and one-third of zinc. It is harder than copper, and can be more easily worked. The addition of from one to two per cent of lead improves the quality of brass for most purposes. The *yellow metal*, used for the sheathing of ships, contains sixty per cent of copper. *Bronze*, *gun-metal*, *bell-metal*, and *speculum-metal* are alloys of copper and tin in varying quantities. They are all remarkable for the property of being hard and brittle when slowly cooled, but of becoming soft and malleable if, when red-hot, they are cooled suddenly by dipping into cold water.

79. *TIN.* — The ores of tin — although this metal has been known from very early times — occur in but few localities, and the metallic tin is not found in nature. The chief European sources of tin are the Cornish mines, where it is found as *tin-stone*, SnO_2. It was probably from these mines that the Phœnicians and Romans obtained all the tin which they employed in the manufacture of bronze. Tin-stone is also met with in Malacca and Borneo and Mexico. In order to prepare the metal, the tin-stone is crushed and washed to remove mechanically the lighter portions of rock with which it is mixed, and the purified ore is then placed in a reverberatory furnace, with anthracite or charcoal, and a small quantity of lime. The oxide is thus reduced, and the liquid metal, together with the slag, consisting of calcic silicate, falls to the lower part of the furnace. The blocks of tin, still impure, are then refined by gradually melting out the pure tin, leaving an impure alloy behind.

Tin has a white color resembling that of silver; it is soft, malleable, and ductile, but has little tenacity. When bent, pure tin emits a peculiar crackling sound. It melts at 442°, and is not sensibly volatile. Tin does not lose its lustre on exposure to the air, whether dry or moist, at ordinary temperatures; but if strongly heated, it takes fire, and a white powder of stannic oxide is formed.

Owing to its brilliancy, and its power of resisting ordinary atmospheric changes, tin is largely used for coating iron, copper, and other metals, which are more abundant and more easily oxidized. *Tin plate*, or *sheet tin*, as it is called, is iron thus coated with tin. The thin sheets of iron are thoroughly cleaned with sulphuric acid, and then immersed in melted tin for an hour or so. Copper is tinned by brushing the melted tin over its surface, which must first be made perfectly clean.

Tin is sometimes used for water-pipes; and there is a

popular impression that it is never acted upon by water. In certain localities, however, it oxidizes rapidly, and is soon rendered worthless. The safety of tin pipes, as compared with lead, does not consist in their exemption from corrosive action, but in the harmlessness of the resultant oxides and carbonates.

80. *Alloys of Tin.*—Several alloys of tin, besides those already mentioned (78), are employed in the arts. *Britannia metal* is an alloy of equal parts of brass, tin, antimony, and bismuth. The best *pewter* is an alloy of four parts of tin to one of lead. Common *solder* contains equal parts of tin and lead, and is more fusible than lead.

81. *ZINC.*—Zinc is an abundant and useful metal, closely resembling magnesium in its chemical character, but it is much more easily extracted from its ores. The chief ores of zinc are the *sulphide* or *blende*, the *carbonate* or *calamine*, and the *red oxide*. In order to extract the metal, the powdered ore is roasted, to convert the sulphide or carbonate into oxide; the roasted ore is then mixed with fine coal or charcoal, and strongly heated in crucibles or retorts of peculiar shape; the zincic oxide is reduced by the carbon, carbonic oxide gas comes off, and the metallic zinc distils over, and is easily condensed.

Zinc is a bluish-white metal, with a crystalline structure. It is brittle at the ordinary temperature, but when heated to between 200° and 300°, it may be rolled out or hammered with ease. If more strongly heated, it is again brittle, and may be broken up in a mortar. Zinc melts at 773°, and at a bright red heat it begins to boil, and volatilizes; or if air be present, it takes fire, and burns with a greenish flame, forming zincic oxide. Zinc is not acted upon by moist or dry air, and hence it is largely used in the form of sheets, and is employed as a protecting covering for iron. The sheets of iron are plunged into melted zinc, covered with sal-ammoniac, which keeps the surface of the zinc free from oxide, and allows the two metals to unite. Iron thus coated with zinc is said to be *galvanized*. *German silver* is an alloy of zinc, copper, and nickel.

82. *IRON SALTS.*—The oxides of iron are four in number: (1) *ferrous oxide*, FeO, from which the green or ferrous salts are derived; (2) *ferric oxide*, Fe_2O_3, yielding the yellow ferric salts; (3) the *magnetic*, or *black oxide*, Fe_3O_4, which does not form any definite salts; (4) *ferric acid*, H_2FeO_4, a weak acid, forming colored salts with potassium. The ferrous oxide colors glass green, and gives the peculiar tint to common bottle-glass. The most important of the ferrous salts are,—

(1) *Ferrous Sulphate*, $FeSO_4$. This soluble salt, sometimes called *green vitriol*, is obtained by dissolving (1) metallic iron, or (2) ferrous sulphide, in sulphuric acid, or by the slow oxidation of *iron pyrites*, FeS_2.

$$(1)\ Fe + H_2SO_4 = FeSO_4 + H_2.$$
$$(2)\ FeS + H_2SO_4 = FeSO_4 + H_2S.$$

The solution thus obtained yields, on evaporation, large green crystals of the salt. It is largely used in the manufacture of several black dyes, and is one of the constituents of writing ink. Like ferrous oxide, this salt easily takes up oxygen, producing a new salt called *ferric sulphate*.

(2) *Ferrous acetate*, formed by the action of acetic acid on iron, is largely used in dyeing and calico-printing.

Of the ferric salts, the *chloride*, Fe_2Cl_6, is the most important; it forms in brilliant red crystals when chlorine gas is passed over heated metallic iron.

83. *LEAD SALTS.*—Three compounds of lead and O are known: (1) *Litharge*, or *plumbous oxide*, PbO, a straw-colored powder obtained by heating lead in a current of air. It is used in painting and in glass-making, and with acids forms the important lead salts. (2) *Plumbic oxide*, PbO_2. (3) *Red oxide*, or *red lead*, a compound of the two last oxides, having the composition, 2 PbO, PbO_2. It is obtained by exposing litharge to the air at a moderate red heat, oxygen being absorbed. Red lead is used as a paint and in glass-making.

Of the soluble salts of lead, the *nitrate*, $Pb2NO_3$, is the most important. This compound is obtained by dissolving the oxide, the carbonate, or metallic lead in warm nitric acid. *Plumbic acetate*, or *sugar of lead*, is also a soluble salt. Almost all the other lead salts are insoluble in water. *Plumbic carbonate*, or *white lead*, $PbCO_3$, is a substance much used in the arts as a paint. The salt may be obtained in the pure state by precipitating a cold solution of the nitrate with an alkaline carbonate, when it falls down as a white powder. For preparing the salt in quantity, two plans are employed: the one similar in principle to that described for the pure salt; and the other an old and interesting process, known as the *Dutch method.* In this process thin sheets of lead are rolled into a coil, and each coil placed in an earthen pot containing a small quantity of crude vinegar, which, however, does not come in contact with the lead. Several hundred of these jars and coils are packed on a floor in a bed of stable manure or spent tan-bark, and then covered with boards, while a second layer of pots similarly charged is placed above, and this is continued until the building is filled. After remaining thus for several weeks, the coils are taken out, when the greater part of the lead is found to be converted into white carbonate. It appears that a lead acetate is first formed, and that the acetic acid is gradually driven out from its combination by the carbonic acid evolved from the putrefying organic matter, and thus enabled to unite with another portion of the lead lying underneath that which was first attacked.

Plumbic sulphide, or *galena*, PbS, is found native, and is the chief ore of the metal. It is prepared as a black precipitate by passing sulphuretted hydrogen gas through a solution of a lead salt. Galena has a bright bluish-white metallic lustre. *Plumbic sulphate* $PbSO_4$, is a white insoluble salt, which is found native, and is prepared artificially by adding sulphuric acid to a soluble lead salt. *Plumbic chloride*, $PbCl_2$, is prepared by adding muriatic acid to a strong solution of plumbic nitrate, when a crystalline precipitate of plumbic chloride is formed. It dissolves in about thirty parts of boiling water, separating out in shining needles on cooling. *Plumbic iodide*, PbI_2, is precipitated in the form of splendid yellow spangles, when hot solutions of potassic iodide and plumbic nitrate are mixed and allowed to cool. *Plumbic chromate*, $PbCrO_4$, is a yellow insoluble salt, generally known as *chrome yellow*, and much used as a paint.

84. *COPPER SALTS.*—Copper forms two oxides, *cuprous oxide*, Cu_2O, and *cupric oxide*, CuO. Cuprous oxide imparts to glass a ruby color. Cupric oxide is soluble in acids, and forms a series of salts. The most important of the soluble salts are *cupric sulphate*, $CuSO_4$, and *cupric acetate*, or *verdigris*. The former is often called *blue vitriol*, and is largely manufactured by dissolving copper or its oxide in sulphuric acid. It is used in calico-printing, and in making *Scheele's green*, or *cupric arseniate*, and other paints.

As many of the vegetable acids act upon copper so as to form poisonous salts, copper or brass vessels should never be used for cooking unless they are lined with tin.

85. *SALTS OF ZINC.* — Zinc forms but one oxide, ZnO, which is much used as a white paint. Its most important salt is *zincic sulphate*, or *white vitriol*, which is used in calico-printing.

86. *SALTS OF TIN.* — Tin forms two oxides, the *stannous*, SnO, and the *stannic*, SnO_2. *Stannous chloride*, $SnCl_2$, formed by dissolving tin in muriatic acid, is termed *tin salts* in commerce; it is largely manufactured for the calico-printer and dyer, who use it as a *mordant;* that is, to *fix* the color in the fibre of the cloth, so that it will not wash out. *Potassic stannate* (*stannate of potash*), K_2SnO_3, formed by boiling stannic oxide with soda, is used in the same way. *Stannic chloride*, $SnCl_4$, is also used by dyers, and is prepared for this purpose by dissolving tin in cold nitro-muriatic acid. Of the sulphides of tin, SnS, *stannous sulphide*, and SnS_2, *stannic sulphide*, are the most important; the former is blackish-gray, and the latter a bright yellow crystalline powder, known as *mosaic gold.*

NOBLE METALS.

The noble metals are *mercury*, *silver*, *gold*, and *platinum*. They are so called because they do not *rust*, that is, combine with the oxygen of the air at ordinary temperatures.

87. *MERCURY.* — The chief ore of mercury is the sulphide, or *cinnabar*, which occurs at Almaden in Spain, at Idria in Illyria, in California, and also in China and

Japan. The metal is easily obtained by roasting the ore, when the sulphur burns off, and the metal volatilizes, and its vapor is condensed in earthen pipes. Mercury is the only metal liquid at the ordinary temperature; it freezes at $-39°$; in the solid state it is malleable and possesses a density of 14.4. It boils at 662°, and gives off a slight amount of vapor at the ordinary temperature. Mercury, when pure, does not tarnish in moist or dry air; but when heated above 700°, it rapidly absorbs oxygen; and it combines directly with chlorine, bromine, iodine, and sulphur. Mercury is largely used in extracting gold and silver from their ores, and for many other purposes.

Mercury dissolves many of the metals, forming *amalgams*. The most important of these is the amalgam of tin and mercury, used for silvering mirrors. "In order to apply it to the glass, a sheet of tinfoil is spread evenly upon a smooth slab of stone, which forms the top of a table carefully levelled, and surrounded by a groove, for the reception of the superfluous mercury. Clean mercury is poured upon the tinfoil, and spread uniformly over it with a roll of flannel; more mercury is then poured on till it forms a fluid layer of the thickness of about half a crown; the surface is cleared of impurities by passing a linen cloth lightly over it; the plate of glass is carefully dried, and its edge being made to dip below the surface of the mercury, is pushed forward cautiously; all bubbles of air are thus excluded as it glides over and adheres to the surface of the amalgam. The plate is then covered with flannel, weights are placed upon the glass, and the stone is gently inclined so as to allow the excess of mercury to drain off; at the end of twenty-four hours it is placed upon a wooden table, the inclination of which is increased from day to day until the mirror assumes a vertical position: in about a month it is sufficiently drained to allow the mirror to be framed."—MILLER.

88. *SILVER.*—Silver was known to the ancients. It is found in the native state, as well as combined with sulphur, antimony, chlorine, and bromine. It is also contained in small quantities in galena, and it can be extracted with profit from the lead prepared from this ore, even when the lead contains only two or three ounces of silver to the ton. The method thus adopted for the extraction of the silver depends upon the fact that the whole of the silver can be concentrated into a small portion of lead by crystallization; metallic lead free from silver separates out in crystals, and a rich alloy is left. When this reaches the concentration of 300 oz. of silver to the ton, the alloy undergoes the operation of *cupellation*, in which the mixture is melted in a furnace on a porous bed of bone-earth, and a blast of air blown over the surface. The lead oxidizes, and the oxide (litharge) fuses, and partly runs away and partly sinks into the porous bed of the furnace, while the silver remains behind in the metallic state.

For the extraction of silver from the other ores, a process termed *amalgamation* is employed, in which mercury is used to dissolve the metallic silver. The argentiferous ores of Germany, in which the silver occurs in combination with sulphur, are roasted in a furnace with common salt, by which means the sulphide is converted into chloride; the mixture is then placed in casks made to revolve, and scrap-iron and water are added. The iron reduces the silver to the metallic state, and when this is fully accomplished, metallic mercury is added; this forms a liquid amalgam with the silver (and gold, if any be present), and, by distilling the mercury off, the silver is obtained in an impure state. Silver has a bright white color and a brilliant lustre, which it does not lose in pure air at any temperature. When melted in the air, it possesses the singular power of absorbing mechanically a large volume (twenty-two times its bulk) of oxygen; this gas it again gives out on solidifying.

Silver is probably the best conductor of heat and electricity known, and is extremely ductile. Sulphur combines at once with silver, forming a black sulphide; silver articles long exposed to the air tarnish from this cause. It is because they contain a small quantity of sulphur that eggs, mustard, and horseradish blacken silver spoons. Where coal-gas is used, silver articles not unfrequently tarnish, owing to the fact that the sulphuretted hydrogen has not been thoroughly removed in the purification of the gas.

89. *GOLD.*—Gold is usually found in the metallic state. It occurs in veins in the older rocks, and in the sands of most rivers, and, although generally found in small quantities, it is a widely diffused metal. In order to obtain the gold, the *detritus* or sand which contains the metal is washed in a *cradle* or other arrangement, by means of which the lighter particles of mud or mineral are washed away, while the heavier grains of gold sink to the bottom of the vessel. When gold has to be worked in the solid rock, the mineral is crushed to powder and then shaken up with mercury, and the gold thus extracted by amalgamation.

Gold has a brilliant yellow color, and, in thin films, transmits green light; it is nearly as soft as lead; it can be drawn out into fine wire, and is the most malleable of all the metals. It does not tarnish at any temperature, in dry or moist air; nor is it, like silver, affected by sulphur. It is not acted upon by any single acid (except *selenic*), but dissolves in presence of free chlorine and in nitro-muriatic acid. At high temperatures, gold is slightly volatile. The gold used for coin is an alloy of gold and copper, in the proportion of 11 parts of gold to

1 of copper. This alloy is harder and more fusible, but less ductile, than pure gold. The purity of gold is commonly expressed in *carats*. Pure gold is said to be 24 *carats fine;* an alloy containing $\frac{23}{24}$ of pure gold is called 23 *carats fine;* $\frac{18}{24}$ of pure gold (the finest usually employed for jewelry), 18 *carats fine;* and so on.

90. *PLATINUM.* — Platinum is a comparatively rare metal, which always occurs in the native state, and generally alloyed with five other metals; namely, *palladium*, *rhodium*, *iridium*, *osmium*, and *ruthenium*. This alloy occurs in small grains in detritus and gravel in Siberia and Brazil.

The original mode of obtaining the metal was to dissolve the ore in aqua-regia (nitro-muriatic acid), and precipitate the platinum (together with several of the accompanying metals) with sal-ammoniac, as the insoluble double chloride of ammonium and platinum, $2NH_4Cl,PtCl_4$. This precipitate, when heated, yields metallic platinum in a finely divided or spongy state; and this sponge, if forcibly pressed and hammered when hot, gradually becomes a coherent metallic mass, the particles of platinum welding together, when hot, like iron. A new mode of preparing the metal has recently been proposed, the ore being melted in a very powerful furnace, heated with the oxy-hydrogen blow-pipe. In this way a pure alloy of platinum, iridium, and rhodium is formed, the other constituents and impurities of the ore being either volatilized by the intense heat, or absorbed by the lime of which the crucible is composed. This alloy is in many respects more useful than pure platinum, being harder and less easily attacked by acids.

Platinum has a bright white color, and does not tarnish under any circumstances in the air. It is extremely infusible, and can only be melted by the heat of the oxy-hydrogen blowpipe. It dissolves in aqua-regia, but is not acted upon by the ordinary acids, and hence platinum vessels are much used in the laboratory. Caustic alkalies, however, act upon the metal at high temperatures. When finely divided, metallic platinum has the power of condensing gases upon its surface in a remarkable degree. When a mixture of oxygen and hydrogen is brought in contact with spongy platinum, the heat developed by the condensation of the gases is sufficient to ignite them.

91. *SALTS OF MERCURY.* — Mercury forms two oxides, the *mercurous*, Hg_2O, and the *mercuric*, HgO, and two corresponding classes of salts. The *mercuric sulphide*, HgS, is known as *cinnabar*, or *vermilion*, much used as a paint. *Mercurous chloride*, Hg_2Cl_2, or *calomel*, and *mercuric chloride*, $HgCl_2$, or *corrosive sublimate*, are used in medicine. The bromides and iodides of mercury are used in photography.

92. *SALTS OF SILVER.* — Silver forms two oxides, Ag_4O, *argentic suboxide*, and Ag_2O, *argentic oxide*. The latter is the base of the ordinary silver salts. *Argentic nitrate*, $AgNO_3$, is the most important soluble salt of silver. It is obtained in the form of large transparent tabular crystals, on evaporating a solution of silver in nitric acid. It fuses easily on heating; and when cast into sticks, goes by the name of *lunar caustic*. This salt undergoes decomposition when exposed to the sunlight in contact with organic matter, and a black substance, probably the suboxide, is formed; hence it is employed in the manufacture of *indelible ink* for marking linen and other fabrics.

Of the insoluble silver salts, the *chloride*, $AgCl$, is the most important. This salt occurs in nature, and is then known as *horn silver*. It is precipitated as a white curdy mass, when a solution of a chloride and a silver salt are brought together. When exposed to the light, the white chloride becomes tinted of a purple color, which increases in shade as the action of light continues. This coloration arises from a partial decomposition of the salt, a small quantity of subchloride and free muriatic acid being formed. In presence of organic matter this change takes place much more rapidly, and upon this fact the art of *photography* depends. Argentic chloride is easily soluble in a solution of sodic hyposulphite, and it is for this reason that the latter salt is used for *fixing* photographic pictures, that is, dissolving out the unaltered silver salt, and thus rendering the image permanent. *Argentic bromide*, AgBr, is also acted upon by the light, and is soluble in ammonia and an alkaline hyposulphite. The *iodide*, AgI, is a yellow powder, insoluble in water and ammonia, but dissolved by an alkaline hyposulphite.

The *cyanide*, AgCN, as well as other salts of silver, is used in electro-plating.

93. *SALTS OF GOLD.*—Gold forms two oxides, the *aurous*, Au_2O, and the *auric*, Au_2O_3. Neither of these oxides forms salts with acids, but the latter unites with bases to form salts called *aurates*. *Auric chloride*, $AuCl_3$, obtained by dissolving gold in aqua-regia, and *auric cyanide*, $Au(CN)_3$, are the most important salts of gold, being used in electro-gilding and photography.

94. *SALTS OF PLATINUM.*—Platinum forms two oxides, the *platinous*, PtO, and the *platinic*, PtO_2. *Platinic chloride*, $PtCl_4$, is the most important platinum compound. It is obtained as a yellowish-red solution, by dissolving the metal in aqua-regia. It is used in electro-platinizing.

LESS USEFUL METALS.

The less useful metals are *sodium*, *magnesium*, *aluminium*, *antimony*, *bismuth*, and *nickel*. These are little used in a metallic state, though their compounds are of the highest importance.

95. *SODIUM.*—This metal was discovered by Sir H. Davy, by the decomposition of soda with the galvanic current. It can be procured by reducing the carbonate in presence of carbon, and is now manufactured in considerable quantities for the preparation of other metals, especially magnesium and aluminium. Sodium is a silver-white metal, soft at ordinary temperatures, and melting at 207°; it volatilizes below a red heat, yielding a colorless vapor. The compounds of sodium are very widely diffused, being contained in every speck of dust. They exist in enormous quantities in the older rocks; but they are most readily obtained from sea-water, which contains nearly 3 per cent of sodic chloride (common salt), or from the large deposits of this substance which occur in the form of rock-salt.

96. *MAGNESIUM.*—This metal occurs in large quantities in combination with lime and carbonic acid, in *dolomite* or mountain limestone, and also in sea-water and certain mineral springs, as chloride and sulphate. The metal itself has only recently been prepared in quantity. It is best obtained by heating magnesic chloride with metallic sodium, sodic chloride and metallic magnesium being formed. This metal is of a silver-white color, and fuses at a low red heat. It is volatile, and may be easily distilled at a bright red heat. When soft, it can be pressed into wire, and with care it may be

cast like brass; although, when strongly heated in the air, it takes fire, and burns with a dazzling white light, with the formation of its only oxide, *magnesia*, MgO. The light emitted by burning magnesium wire is distinguished for its richness in chemically active rays, and is therefore employed as a substitute for sunlight in photography. It is coming to be used for purposes of illumination.

97. *ALUMINIUM.*—This metal occurs in large quantities, combined with silicon and oxygen, in *felspar* and all the older rocks, and also in *clay*, *marl*, *slate*, and in many crystalline minerals. Metallic aluminium is obtained by passing the vapor of aluminic chloride over metallic sodium. It has recently been manufactured on a large scale, both in England and France, and from its lightness (specific gravity 2.6), its tenacity, and its bright lustre, it has been used for optical purposes as well as for ornamental work. Some of its alloys, as *aluminium bronze*, promise to become of great value in the arts.

98. *ANTIMONY.*—Metallic antimony occurs native, but its chief ore is the *trisulphide*, Sb_2S_3. The metal is easily reduced, by heating the sulphide with about half its weight of metallic iron, when ferrous sulphide and metallic antimony are formed. Antimony may also be reduced by mixing the ore with coal, and heating in a reverberatory furnace. Antimony is a bright bluish-white metal. It is very brittle, and can readily be powdered in a mortar. It melts at about a red heat, and may be distilled at a white heat, in an atmosphere of hydrogen. Antimony undergoes no alteration in the air at ordinary temperatures, but rapidly oxidizes if exposed to air when melted; and if heated more strongly, it takes fire, and burns with a white flame, giving off dense white fumes of

oxide. Antimony is not attacked either by dilute muriatic or sulphuric acid; but nitric and nitro-muriatic acid dissolve it easily. The alloys of antimony are largely used in the arts. Of these, *type-metal* (an alloy of lead and antimony) is the most important; it contains from 17 to 20 per cent of the latter metal.

99. *BISMUTH.*—This metal is found in small quantities in the native state, but occurs more often as a sulphide; it is easily reduced to the metallic state, and then exhibits a pinkish-white color. It melts at 507°, and is volatilized at a white heat. Bismuth does not oxidize in dry air at the ordinary temperature; but if heated strongly, it burns with a blue flame, forming an oxide; it also takes fire when thrown into chlorine gas. Bismuth dissolves easily in nitric acid. The metal is chiefly used as an ingredient of *fusible metal.* The alloy of 2 parts of bismuth, 1 of lead, and 1 of tin, melts at 201°; that of 8 parts of bismuth, 5 of lead, and 3 of tin, melts at 212°. The compounds of bismuth are used in the porcelain manufacture, in medicine, and as cosmetics.

Both bismuth and antimony closely resemble arsenic in the compounds which they form. Thus we have Sb_2O_3, Sb_2O_5, and H_3Sb; Bi_2O_3 and Bi_2O_5 (67). Hence bismuth and antimony are now generally classed among the *non-metallic* elements of the *nitrogen group.*

100. *NICKEL.*—Nickel occurs in large quantities combined with arsenic, as *kupfernickel;* also together with cobalt in *speiss;* and it is now prepared in considerable quantities for the manufacture of *German silver*, an alloy of nickel, zinc, and copper. Nickel is a white, malleable, and tenacious metal; it melts at a somewhat lower temperature than iron, and is strongly magnetic, but loses this property when heated to above 600°.

101. *SALTS OF SODIUM.*—There are two compounds of sodium and oxygen, Na_2O, *sodic oxide*, and Na_2O_2, *sodic peroxide*. Sodic oxide has a strong affinity for water, with which it combines to form HNaO, *sodic hydrate* (*caustic soda*). This is a white solid, very soluble in water, very caustic and alkaline, and is largely used in making soap.

Caustic soda is now manufactured on a very large scale, by boiling lime and sodic carbonate (carbonate of soda) with water, and evaporating down the clear solution. The reaction is as follows:—

$$CaO + Na_2CO_3 + H_2O = CaCO_3 + 2HNaO.$$

102. *Sodic Chloride.*—Sodic chloride, NaCl, is *common salt*. It is from this compound that nearly all the other sodic compounds are prepared.

In Southern Europe, on the shores of the Mediterranean, large quantities of salt are obtained from sea-water. The water is allowed to flow into large shallow pools, called *salt-pans*, where it is evaporated by the agency of the air and the sun. Only pure water passes off in the form of vapor, and the solution grows more and more concentrated. After a time, the brine is pumped into large iron pans, and evaporated by artificial heat until the salt crystallizes.

In many parts of the earth there are salt lakes, often called *seas*. The streams flowing into them are more or less impregnated with salt, which they dissolve from the soil over which they flow. Such lakes are natural salt-pans, for the water which is evaporated from them is pure, and they consequently become more and more saturated with salt, till it is finally deposited on the bottom in solid crystals. The most remarkable of these lakes is the Great Salt Lake in Utah.

Such salt lakes appear to have existed in all geological

ages, and by long-continued evaporation to have given rise to large masses of solid salt, called *rock salt*. These beds of rock salt are found in all parts of the earth. The most remarkable are at Wieliczka in Poland, and at Cardona in Spain. The former is 500 miles long, 20 miles broad, and 1,200 feet deep. It has been worked for several centuries. Some of the galleries excavated in it are 30 miles long. The bed at Cardona is a "mountain of salt," 400 or 500 feet high, and the salt is of the greatest purity.

Salt is obtained from these beds sometimes by excavation and sometimes by solution. In the latter case, a hole is sometimes bored down to the bed of salt, often to the depth of 1,200 or 1,500 feet. A tube somewhat smaller than the bore is then introduced, and fresh water allowed to flow down outside the tube. This water, on reaching the bed, becomes saturated with salt, and rises in the tube. Since, however, it is heavier than fresh water, it will not rise to a level with the latter on the outside of the tube. If the bore is 1,200 feet deep, the brine will rise within about 200 feet of the top. It must be pumped up the remaining distance. The solution is then evaporated, and solid salt obtained.

It not unfrequently happens that natural springs occur in such localities that their waters come into contact with beds of salt. Such springs are called *salt springs*, or *brine springs*. They sometimes come to the surface; but oftener they can only be reached by boring, and the brine brought to the surface by means of the pump. The water of these natural springs is usually less concentrated than that of the artificial springs described above. In England and in this country the salt produced is nearly all obtained from brine springs.

The most important salt springs of this country are in Central New York, in Virginia, and in Pennsylvania.

About 6,000,000 bushels of salt are produced annually in New York. When the brine is not a saturated solution, it is usually concentrated by partially evaporating it in large shallow pans by the air and the sun. It is then further concentrated and crystallized by artificial heat.

103. *Sodic Carbonate.* — Sodic carbonate, Na_2CO_3, is known in commerce as *soda-ash*, and is manufactured on an enormous scale. It is largely used in glass-making, soap-making, bleaching, and many other processes in the arts.

Before the French Revolution, the Continental nations of Europe derived their soda-ash mainly from Spain, where it was made from the ashes of certain marine plants, which contain a large amount of soda. The soda-ash thus obtained was called *barilla*.

In England, the soda-ash used was mainly obtained from the ashes of a sea-weed called *kelp*, which grows abundantly on the north and west coast of Ireland, and on the west coast and the islands of Scotland.

One of the first effects of the war of the French Revolution was to cut off the supply of alkali from Spain. About this time, a French chemist, Le Blanc, discovered a process by which sodic carbonate, or soda-ash, could be obtained from common salt, or *sodic chloride*, NaCl. This process became publicly known through a commission appointed, during the first year of the Republic, to investigate the subject of alkali manufacture.

Le Blanc's method is the same as that now used in the making of sodic carbonate, which has come to be one of the most important branches of chemical manufacture. The process may be divided into two stages: —

(1) Manufacture of sodic sulphate, or *salt-cake*, from sodic chloride; called the *salt-cake process*.

(2) Manufacture of sodic carbonate, or soda-ash, from salt-cake; called the *soda-ash process*.

(1) *Salt-cake Process.* — This consists in the decomposition of salt, by means of sulphuric acid. This is effected in a furnace, called the *salt-cake furnace*, a section of which is represented in Figure 17. It consists

Fig. 17.

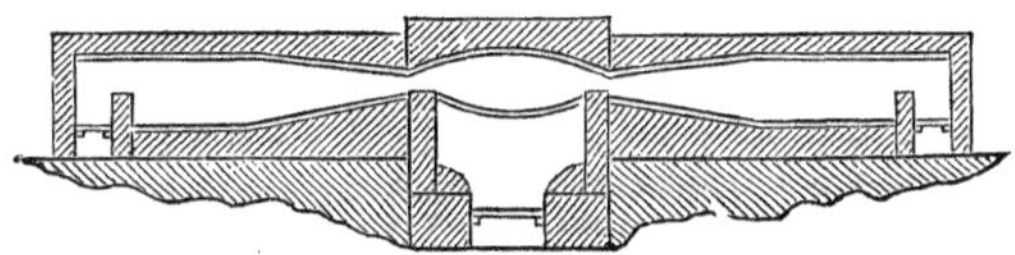

of a large covered iron pan, placed in the centre of the furnace, and heated by a fire placed underneath; and two *roasters*, or reverberatory furnaces, placed one at each side, on the hearths of which the salt is completely decomposed. The charge of half a ton of salt is first placed in the iron pan, and the sulphuric acid allowed to run in upon it. Hydric chloride (muriatic acid) is evolved, and escapes through a flue with the products of combustion into towers, or *scrubbers*, filled with coke or bricks moistened with a stream of water; the acid vapors are thus condensed, while the smoke and heated air pass up the chimney. The reaction is as follows: —

$$2NaCl + H_2SO_4 = Na_2SO_4 + 2HCl.$$

After the mixture of salt and acid has been heated for some time in the iron pan, and has become solid, it is raked out upon the hearths of the *roasters*, where the flame and heated air of the fire complete the decomposition into sodic sulphate and hydric chloride.

(2) *Soda-ash Process.* — This consists (1) in the preparation of the sodic carbonate, and (2) in the separation and purification of the same. The first chemical change which the salt-cake undergoes is its reduction to sodic *sulphide*, by heating it with powdered coal: —

$$Na_2SO_4 + C_4 = Na_2S + 4CO.$$

The second decomposition is the conversion of the sodic sulphide into sodic carbonate, by heating it with chalk or limestone (calcic carbonate): —

$$Na_2S + CaCO_3 = Na_2CO_3 + CaS.$$

These two reactions are, in practice, carried on at once; a mixture of 10 parts of salt-cake, 10 parts of chalk or limestone, and 7.5 parts of coal, being heated in a reverberatory furnace, called the *balling furnace*, until it fuses, and the decomposition is complete, when it is raked out into iron wheelbarrows to cool. This process is generally called the *black-ash process*, from the color of the fused mass.

The next operation consists in the separation of the sodic carbonate from the insoluble calcic sulphide and other impurities. This is easily accomplished by *lixiviation*, or dissolving the former salt out in water. On evaporating down the solution, for which the waste heat of the balling furnace is used, and calcining the residue, the soda-ash of commerce is obtained.

No less than 200,000 tons of common salt are annually consumed in the alkali works of Great Britain, for the preparation of nearly the same weight of soda-ash, the value of which is about $10,000,000.

104. *Bicarbonate of Soda.* — The salt known as *bicarbonate of soda*, $HNaCO_3$, is obtained by exposing the crystallized sodic carbonate in an atmosphere of carbonic acid. It is a white crystalline powder, used in medicine and for making effervescing drinks; also in bread-making, as a substitute for yeast.

105. *Other Sodic Salts.* — *Sodic nitrate* (*soda saltpetre*), $NaNO_3$, is found in northern Chili in large beds. It is used in making nitric acid, and as a manure.

Sodic sulphate (*sulphate of soda*), Na_2SO_4, is familiarly known as *Glauber's salts*.

Sodic hyposulphite (*hyposulphite of soda*) is used in photography (92).

Sodic borate, or *borax*, is considerably used in the arts, especially as a *flux* in soldering. It dissolves the oxide formed upon the surfaces to be united, and thus keeps them clean and bright.

Sodic silicate (*silicate of soda*) is called *soluble glass*, and is used to some extent as a paint or varnish for fixing fresco colors, and for preserving stone-work, and quite largely in calico printing and dyeing.

106. *SALTS OF MAGNESIUM.* — Magnesium has but one oxide, MgO, or *magnesia*. The most important salt of this oxide is the *magnesic sulphate* or *Epsom salts*, used in medicine.

107. *SALTS OF ALUMINIUM.* — The only oxide of aluminium is *alumina*, Al_2O_3. It occurs native in a nearly pure and crystalline state as *corundum*, *ruby*, *sapphire*, and *emery*. Alumina is largely used as a *mordant* in dyeing and calico-printing, as it has the power of forming insoluble compounds, called *lakes*, with vegetable coloring-matter, and thus fixes the color in the pores of the cloth so that it cannot be washed out (85).

The soluble aluminic sulphate, Al_23SO_4, is prepared on a large scale for the use of the dyer, by decomposing clay by the action of sulphuric acid. The solid mixture of silica and aluminic sulphate thus obtained goes by the name of *alum-cake*. The most useful compounds of alumina are, however, the *alums*, a series of double salts, which aluminic sulphate forms with the alkaline sulphates. *Common potash alum* has the composition $KAl2SO_4$. It may be prepared by dissolving the two sulphates together, and allowing the compound salt to crystallize, but it is usually obtained from the decomposition of a shale or clay containing iron pyrites, FeS_2. This substance gradually undergoes oxidation when the shale is roasted (taking oxygen from the air), producing sulphuric acid, which unites with the alumina of the clay, and, on the addition of a potassic compound, alum crystallizes out. A salt called *ammonium alum*, and containing H_4N, instead of K, is at present prepared on a large scale; the ammonia liquor of the gas-works, together with sulphuric acid, being added to the burnt shale, instead of a potash salt.

There are a great many other alums known, in which Fe_2O_3, Cr_2O_3, or Mn_2O_3, *oxides* of *iron*, *chromium*, and *manganese*, are substituted for the alumina in common alum. *Clay* is an *aluminic silicate* resulting from the disintegration and decomposition of felspar by the action of air and water. *Kaolin* or porcelain clay is the purest form of clay, containing no iron or other impurities. There are many very beautifully crystalline minerals, as *garnet* and *mica*, consisting of aluminic silicates combined with silicates of other metals.

108. *Glass, Porcelain, and Earthen-ware.* — The silicates of the alkalies are soluble in water and non-crystalline; those of the *alkaline earths* (Ba, Sr, and Ca), are soluble in acid and crystalline; while compounds of the two are insoluble in water and acids, and do not assume a crystalline form. Such a compound, when fused, is termed a *glass*. There are four different kinds of glass used in the arts, differing in their chemical composition, and exhibiting corresponding differences in their properties: (1) *Crown* or *window* and *plate glass*, composed of silicates of soda and lime; (2) *Bohemian glass*, consisting of silicates of potash and lime; (3) *flint glass* or *crystal*, containing silicates of potash and lead; and (4) common green *bottle-glass*, composed of silicates of soda, lime, iron, and alumina. The first and third of these

kinds of glass are easily fusible, while the second or potash glass is much less fusible; the addition of plumbic oxide increases the specific gravity and lustre of the glass as well as its fusibility. The common glass articles of household use are generally made of flint glass, while for chemical apparatus a soda-lime glass is to be preferred. The potash-lime glass is much employed where an infusible or hard glass is needed. The fourth kind of glass is an impure mixture of various silicates, employed for purposes in which the colors and fineness of the glass are not of consequence.

In the preparation of all the fine qualities of glass, great care is requisite in the selection of pure materials, as well as in the processes of manufacture. Generally the materials are melted together with a quarter to a half their weight of *cullet*, or broken glass of the same kind. After the glass articles have been blown or cast, they must be *annealed*, that is, made less brittle by being very slowly heated and as slowly cooled in an oven arranged for the purpose.

Certain metallic oxides have the power of *coloring* glass when they are added in small quantity. Thus ferrous oxide produces a deep green color (bottle-glass), while the oxides of manganese impart a purple tint to glass. Advantage is taken of this in the preparation of colorless glass, for as it is difficult to obtain materials perfectly free from iron, which imparts a green color, a small quantity of black oxide of manganese is added to the mixture, and the violet color thus produced neutralizes the green, and a nearly colorless glass is the result. The addition of arsenious oxide effects the same end by oxidizing the ferrous to ferric oxide. The colors of precious stones are imitated by adding certain oxides to a brilliant lead glass called *paste;* thus the blue of the sapphire is given by a small quantity of cobaltic oxide, while cuprous oxide imparts a ruby-red color, and ferric oxide a yellow color resembling topaz.

The various forms of *porcelain* and *earthen-ware* consist of aluminic silicate or clay in a more or less pure state, covered with some substance which fuses at a high temperature, and forms a *glaze*, giving a smooth surface and binding the material closely together, and thus filling up the pores of the baked clay. For the manufacture of porcelain the finest *kaolin* or China clay is used, while for common earthen-ware an inferior clay may be employed. The glaze used for porcelain is generally finely powdered felspar, the *biscuit*, or porous ware, being first dipped into a vessel containing this substance suspended in water and then strongly heated. The articles thus coated can be used for chemical purposes, as this glaze withstands the action of acids. For earthen-ware the so-called *salt glaze* is used; obtained by throwing some common salt into the furnaces containing the strongly heated ware. The salt is volatilized and undergoes decomposition on the heated surface, depositing a fusible silicate upon it, and rendering the ware impervious to moisture.

109. *SALTS OF ANTIMONY.* — Antimony has two oxides, Sb_2O_3, and Sb_2O_5, and two corresponding sulphides and chlorides. The most important of its salts is *tartar-emetic*, or *tartrate of antimony and potash*, used in medicine.

110. *SALTS OF BISMUTH.* — Bismuth forms two oxides, analogous to those of antimony. The most important of the salts is the *nitrate*, $Bi3NO_3$, which is easily procured by dissolving the metal in nitric acid.

111. *SALTS OF NICKEL.* — There are two

oxides of nickel, NiO, and Ni_2O_3. The former of these gives rise to salts which have a peculiar apple-green color. The latter is a black powder, prepared by adding a solution of bleaching-powder to a soluble nickel salt. The most important soluble salt is the *nitrate*, $Ni2NO_3$.

METALS NOT USED IN A FREE STATE, BUT VALUABLE FOR THEIR SALTS.

There are certain metals which are not used in a free state, but which form salts of considerable importance. These metals are *potassium*, *calcium*, *strontium*, *barium*, *manganese*, *chromium*, and *cobalt*.

112. *POTASSIUM.* — This metal was discovered in the year 1807, by Sir Humphrey Davy, who decomposed potash by means of a powerful galvanic current. Before this time the alkalies and alkaline earths were supposed to be elementary bodies. The metal is now prepared by heating together potash and carbon to a high temperature in an iron retort.

Potassium is a bright, silver-white metal, which can be easily cut with a knife at the ordinary temperature; it is brittle at 32°, and melts at 144°. It rapidly absorbs oxygen when exposed to the air, and becomes converted into a white oxide.

The original source of potassic compounds is in the felspar of the granitic rocks, as these contain from .02 to .03 of this metal. Up to the present time, no cheap and easy mode has been found of separating the potash from the silicic acid with which it is combined in felspar. Plants, however, are able slowly to separate out and assimilate the potash from these rocks and soils; so that, by burning the plant and leaching the ashes with water, soluble potash-salt is obtained.

This is the crude potassic carbonate, called, when purified by crystallization, *pearlash*. It is the substance from which most of the potassic compounds are obtained.

The most important compounds of potassium are *caustic potash*, HKO, *potassic carbonate* (*carbonate of potash*), K_2CO_3, and *potassic nitrate*, or *saltpetre*, KNO_3. This last compound readily parts with its oxygen, and is therefore extensively used in making *gunpowder* and fireworks.

Gunpowder consists of an intimate mixture of saltpetre, charcoal, and sulphur. When the gunpowder is ignited, the sulphur and charcoal burn at the expense of the oxygen in the saltpetre, while the nitrogen is set free. It is to the expansive force of the large volume of gases suddenly set free in a confined space, that the destructive power of gunpowder is due.

Potassic chlorate, $KClO_3$, gives up its oxygen even more readily than the nitrate, and is used, as we have seen, in the preparation of oxygen (10). It is also used in making fireworks and percussion caps.

113. *CALCIUM.* — Calcium forms a considerable portion of the rocks of which the earth is composed, and occurs in very large quantities, forming whole mountain chains of *chalk*, *gypsum*, and *limestone*. The metal is obtained by the decomposition of the chloride by the electric current, or by heating the iodide with sodium. It is a light, yellow metal, which easily oxidizes in the air. When heated in air, it burns with a bright light, forming *lime*, CaO, the only oxide of calcium.

Lime is prepared on a large scale for building and other purposes, by heating limestone (the carbonate) in kilns by means of coal mixed with the stone; the carbonic acid escapes, and *quicklime* or *caustic lime* remains. Pure lime is a white, infusible substance, which combines with water very readily, giving off great heat, and falling to a white powder called *calcic hydrate*, or *slaked* lime. The hydrate is slightly soluble in water, 1 part of it dissolving in 730 parts of cold, but only in 1,300 parts of boiling water, and forming lime-water, which, like the hydrate, has a great power of absorbing carbonic acid from the air. It is indeed partly owing to this property that the hardening or *setting* of mortars and cements made from lime is due. *Mortar* consists of a mixture of slaked lime and sand. A gradual combination of the lime with the silica occurs, and this helps to harden the mixture. *Hydraulic cement*, which hardens under water, is prepared by carefully heating an impure lime containing clay and silica; a compound silicate of lime and alumina appears to be formed on moistening the powder, which then solidifies, and is not acted upon by water. Lime is largely used in agriculture, its action being, (1) to destroy the excess of vegetable matter contained in the soil; and (2) to liberate the potash for the use of the plants from heavy clay soils by decomposing the silicate.

Calcic carbonate (*carbonate of lime*) $CaCO_3$ occurs most widely diffused, as chalk, limestone, coral, and marble; many of those enormous deposits being made up of the microscopic remains of minute sea-animals. The carbonate is almost insoluble in pure water; but readily dissolves when the water contains carbonic acid, and is deposited again, in crystals, as the gas escapes. In this way enormous masses of crystalline limestone are formed. Water charged with carbonic acid and calcic carbonate makes its way through the roof of limestone caverns, and, as the carbonic acid gradually escapes, the calcic carbonate is deposited in dependent masses, like icicles, termed *stalactites;* while the water, falling on the floor of the cavern before it has parted with all its carbonic acid and dissolved limestone, deposits a fresh portion of the crystalline matter, and thus a new growth, or *stalagmite*, gradually rises to meet the *stalactite* which hangs from the roof. In this way a natural pillar of limestone is formed.

Calcic sulphate, $CaSO_4$, occurs in nature, combined with $2H_2O$, as *gypsum* or *alabaster*. It is soluble in 400 parts of water, and is a very common impurity in spring water, giving rise to a *hardness* which cannot be removed by boiling. Gypsum, when moderately heated, loses its water, and is then called *plaster of Paris;* this, when moistened, takes up two atoms of water again and *sets* to a solid mass, and is therefore much used for making casts and moulds.

Calcic chloride (*chloride of calcium*), $CaCl_2$, is formed when limestone or marble is dissolved in muriatic acid; if the solution be then evaporated, colorless needle-shaped crystals are obtained. When these are dried, they form a porous mass which takes up moisture with great avidity, and is much used for drying gases.

114. *STRONTIUM.*—This element occurs in much smaller quantities than calcium, being found in only a few minerals. It likewise occurs in minute quantities in certain spring waters. The metal has a yellowish-white color, and is prepared by the action of a current of electricity on the fused chloride. It resembles calcium closely in its properties. When heated in the air, it burns, forming the oxide *strontia.*

The native salts of strontium, the *carbonate* and *sulphate*, are insoluble, and serve for the preparation of the remaining salts. The *nitrate*, $Sr2NO_3$, and the *chloride*, $SrCl_2$, are soluble in water. These are the only salts of this metal which are employed in the arts. They are used for the preparation of *red fires*, as the volatile salts of strontium have the power of coloring the flame crimson.

115. *BARIUM.* — Barium compounds occur somewhat more widely dispersed than those of strontium, the two most common barium minerals being the *sulphate*, or *heavy spar*, and the *carbonate*, or *witherite*. The metal barium has not yet been obtained in the coherent state; but the metallic powder may be prepared in the same way as calcium and strontium, which it closely resembles in its properties.

The most important barium compounds are the *chloride* and the *nitrate*, used as tests for sulphuric acid, and for giving a green color to the flame in fireworks.

116. *CHROMIUM.* — Chromium is a substance whose compounds do not occur very widely distributed, or in large quantities. They are, nevertheless, much used in the arts as pigments, many of them having a fine bright color. The chief ore of this metal is a compound of the oxides of chromium and iron, called *chrome ironstone*, found in America, Sweden, and the Shetlands. A compound of chromium, lead, and oxygen, called *plumbic chromate* (*chromate of lead*), is also found. Pure chromium appears to be the most infusible of all the metals, as it cannot be melted at a temperature sufficient to fuse and volatilize platinum.

Chromium forms three oxides: *chromous oxide*, CrO, *chromic oxide*, Cr_2O_3, and *chromic peroxide*, CrO_3, which, dissolved in water, has a strong acid reaction, and may then be called *chromic acid*. It neutralizes bases, and forms yellow or red salts, called *chromates*. The most common of these are known as the *chromate* and *bichromate* of *potash*, K_2CrO_4 and $K_2Cr_2O_7$. The former is a yellow salt, and the latter a red salt, which is made on a large scale, and used in the preparation of the various chrome pigments.

The chief of the insoluble chromates is *plumbic chromate*, $PbCrO_4$, or *chrome yellow*, obtained by precipitating a potassic chromate by a soluble lead salt, and largely used for dyeing and other purposes.

117. *COBALT.*—Cobalt is never found in a free state except in small quantities in meteoric iron. It is almost as infusible as iron. It is of a reddish-gray color, hard, and strongly magnetic. It is not used in a metallic state, but many of its compounds are remarkable for the beauty of their colors, and are used as pigments. It forms two oxides, CoO and Co_2O_3. The *zaffre* of commerce is a very impure oxide of cobalt, mixed with two or three times its weight of sand. It is used for coloring glass. *Thenard's blue*, or artificial *ultramarine*, is a compound of cobaltic and aluminic oxides.

118. *MANGANESE.*—Manganese occurs in nature as an oxide, and it can be obtained, though with difficulty, in the metallic state, by heating the oxide very strongly with charcoal. The metal is of a reddish-white color; it is brittle, and hard enough to scratch glass. It decomposes water at the ordinary temperature with evolution of hydrogen; it cannot be preserved in the air without undergoing oxidation, and must be kept under naptha, or in a sealed tube; it is slightly magnetic, and, like iron, combines with carbon and silicon. Metallic manganese is not used in the arts; but an alloy of this metal and iron is now made on a large scale, and used in the manufacture of steel. Some of its oxides are used for evolving chlorine from muriatic acid, and also for giving glass a purple color.

Manganese forms several well-marked oxides: (1) *man-*

ganous oxide, MnO, a basic body, furnishing a series of salts; (2) *manganic oxide*, Mn_2O_3, which also forms salts, but of a much less stable character; (3) *black oxide*, MnO_2, a neutral substance, occurring as an ore of manganese; (4) and (5) two acid oxides, which give rise to salts known as *manganates* and *permanganates*.

Black oxide of manganese is found in many parts of Europe; but the most productive mines are in Thuringia and Moravia. More than 18,000 tons are used annually in England in the manufacture of bleaching powder.

THE RARER METALS.

The other metallic elements are very rare substances, and as yet are of very little practical importance.

119. *Cadmium* (Cd) is a white metal resembling tin, and is found in small quantities in the ores of zinc. It is remarkable for forming very fusible alloys. For example, an alloy of 15 parts of bismuth, 8 of lead, 4 of tin, and 3 of cadmium, melts at about 140° F. The *sulphide* is a valuable yellow pigment, and the *iodide* is used to some extent in photography.

120. *Iridium* (Ir) is the heaviest of the metals. An alloy of iridium and osmium is very hard, and is used for the points of gold pens. It forms three oxides, which readily pass one into another, giving rise to the variety of tints which characterize the solution of its salts. It is because of these changes of color that the metal is called *iridium*, from *iris*, the rainbow.

121. *Tungsten*, or *wolfranium* (W), forms with steel an alloy of remarkable hardness. WO_3, or *tungstic acid*, combines with sodium to form *sodic tungstate* (*tungstate of soda*), Na_2WO_4, which is used in calico-printing, and in combination with starch to render linen and cotton fabrics incombustible.

122. *Uranium* (U) forms an oxide, UO, which gives a peculiar yellow color to glass, and also renders it *fluorescent.*

The other metals are of too little use or interest to be mentioned here.

SUMMARY.

The metals are seldom found in a free state, but usually combined with other elements as *ores.* Their separation from these ores is called *reduction.*

The most useful metals are iron, lead, copper, tin, and zinc.

The noble metals are gold, silver, mercury, and platinum.

The less useful metals are sodium, magnesium, aluminium, antimony, bismuth, and nickel.

The metals not used in a free state, but valuable for their salts, are potassium, calcium, strontium, barium, manganese, chromium, and cobalt.

Among the rarer metals, cadmium, iridium, tungsten, and uranium are of some interest.

Among the most important *metallic compounds* may be mentioned: salt, soda-ash, soda saltpetre, caustic soda, bicarbonate of soda, pearlash, saltpetre, caustic potash, lime, alum, glass, porcelain, green vitriol, blue vitriol, white vitriol, litharge, red lead, and white lead.

CHEMISTRY OF THE ATMOSPHERE.

COMBUSTION.

123. *Composition of the Atmosphere.*—We live at the bottom of an aerial ocean called the *atmosphere*, which is some fifty miles in depth. This atmosphere contains free oxygen; as may be shown by inverting a jar of it over a jar of nitric oxide. The gases, on mixing, become cherry-red (33).

Burn a piece of phosphorus in a jar of air over water. The jar is filled with dense white fumes, which are soon absorbed by the water, which rises and partially fills the jar. Introduce now some nitric oxide into the jar while still over the water, and the gas shows no trace of red color. The free oxygen then can be removed from the air by burning phosphorus in it. After all the oxygen has combined with the phosphorus, and the compound formed has been absorbed by the water, the jar is found to be about one-fifth filled with water; showing that about one-fifth of the air is oxygen.

The gas remaining in the jar is found to be almost pure nitrogen.

Besides the oxygen and nitrogen, there is a small quantity of *carbonic anhydride* (carbonic acid); as may be shown by passing a large volume of air through lime-water. There is also *watery vapor*, which is continually condensed in the form of rain and dew. A trace of *ammonia*, and of *nitric acid*, is also found; and to these are to be added the exhalations continually rising from the earth. The whole quantity of these minor ingredients, with the exception of watery vapor, does not amount to more than one part in a thousand. The proportions of oxygen and nitrogen are almost invariable, while those of the other ingredients are continually fluctuating.

It must be borne in mind that the atmosphere is not a chemical compound, but merely a *mixture* of these different gases. "Indeed, we may regard the globe as surrounded by at least three separate atmospheres, — one of oxygen, one of nitrogen, and one of aqueous vapor, — all existing simultaneously in the same space, yet each entirely distinct from the other two, and only very slightly influenced by their presence."

The following Tables are from Miller: —

Composition of the Atmosphere.

Oxygen	20.61
Nitrogen	77.95
Carbonic Acid	.04
Watery Vapor (average)	1.40
	100.00

Composition in Tons.

Oxygen	1,233,010	billions	of tons.
Nitrogen	3,994,593	"	"
Carbonic Acid	5,287	"	"
Watery Vapor	54,460	"	"

124. *The Products of the Burning of a Candle or Coal-Gas are Carbonic Acid and Water.* — When a lighted taper is held to a candle, or to a jet of coal-gas, the latter takes fire. In what does its burning consist?

Invert a bottle for a short time over the candle or gas-burner; then remove it, pour in a little lime-water, close the mouth of the bottle with the hand, and shake it. The lime-water becomes milky-white; showing that *carbonic acid* has been produced by the burning.

If the candle or gas be burnt under a tin funnel connected with a long glass tube (which must be kept cold), moisture will collect on the inside, and, after a short time, will trickle down, and drop from the end of the tube. When potassium is put into water, it burns with a rose-colored flame. If the liquid from the tube be tested with a bit of potassium, it proves to be water, which is therefore another product of the burning.

Fig. 18.

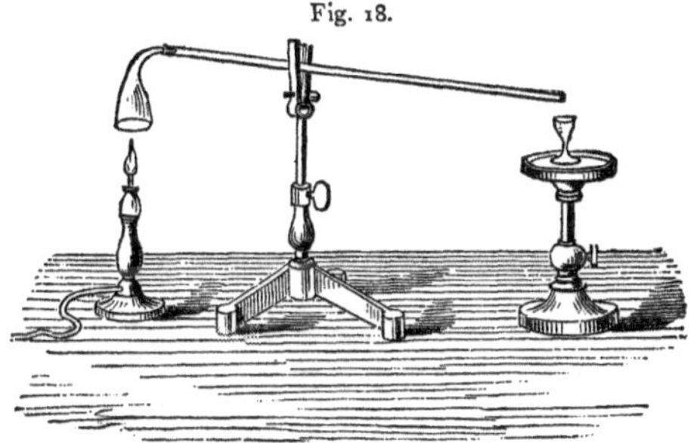

No other substance is produced, except in very minute quantities, by the burning of the candle or the gas.

125. *The Candle, in Burning, removes Oxygen from the Air.*—Arrange a candle, so that it can be covered with a bell-jar, over the water-trough; then light it, and cover it with the jar. It soon ceases to burn, and the water rises in the jar. The burning, then, removes something from the air.

Put a lighted candle into air from which the oxygen has been removed, and it is at once extinguished; showing that it cannot burn without oxygen. It must then be *oxygen* which it removes from the air when it burns.

126. *All ordinary Combustion consists in the Combination of the Oxygen of the Air with the burning Substance.* — We have seen that carbonic acid, CO_2, and water, H_2O, are the products of the burning of a candle. Of the three elements in these products, the O, as we have seen, comes from the air; the C and H exist in the candle.

The force, then, which causes the candle to burn, is *affinity*, and the burning is the combination of the oxygen of the air with the elements of the candle.

Experiments similar to those which we have tried with the candle, will show that any ordinary combustible substance removes oxygen from the air when it burns, and that it cannot burn without oxygen. We conclude, then, that all ordinary combustion is the combination of the oxygen of the air with the burning body.

127. *Why a Draft is necessary in Stoves and Furnaces.* — We see now why our stoves and furnaces must have a draft. As fast as the oxygen is taken from the air by the burning fuel, this air must be removed, and a fresh supply must take its place. In other words, a stream of air must be kept constantly flowing over or through the fuel.

We see, also, how the fire can be regulated by means of the draft. If the doors or dampers through which the air is admitted be partially closed, the supply of air will be diminished, and the burning will therefore be retarded.

128. *Combustibles and Supporters of Combustion.* — Any substance, as coal-gas, which can be made to burn, is called a *combustible;* while any substance, as air or oxygen, in which it can burn, is called a *supporter of combustion.* These terms are convenient, though, strictly speaking, the one substance is no more a combustible or a supporter of combustion than the other. Since the

burning of coal-gas consists in the combination of the gas with oxygen, the oxygen in reality burns, as well as the gas; and, on the other hand, the gas is as much a supporter of the combustion as the oxygen. The burning must of course take place where the gases come together. A jet of oxygen would appear to burn in an atmosphere of coal-gas, just as a jet of coal-gas appears to burn in an atmosphere of oxygen.

Fig. 19

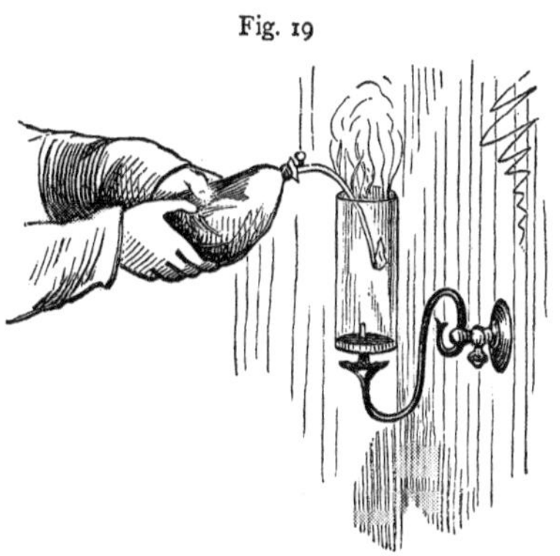

Fit a cork to one end of a lamp chimney, and let the tip of a gas-burner pass through it, as represented in the figure. Allow the gas to escape for some time, and then light it at the top of the chimney. It will burn quietly, and the chimney will evidently be filled with coal-gas. Fill a gas-bag with oxygen, and fasten to the bag a bent glass tube drawn out into a fine jet. Force the oxygen through the tube in a gentle stream, and introduce the end of the tube through the flame into the chimney. As it passes the flame, the oxygen takes fire, and burns brightly in the coal-gas; the oxygen apparently becoming the combustible body, and the coal-gas the supporter of combustion.

In both the flames which we have here, it will be seen that gases are burning where they come together; — the coal-gas and the oxygen of the air, where they meet at the top of the chimney; the oxygen from the bag and the coal-gas, where they meet at the end of the tube inside the chimney.

129. *Oxygen must be Heated before it will combine with ordinary Combustibles.* — The jet of coal-gas has no disposition to burn until a lighted taper is applied to it. The oxygen of the air is at all times in contact with wood and coal, yet they do not burn unless they are first kindled. When even as inflammable a gas as hydrogen is mixed with oxygen, it does not burn unless ignited with a taper or an electric spark. "At the ordinary temperature of the air its chemical affinities are dormant; and, although endowed with forces which are irresistible when in action, it awaits the necessary conditions to call them forth. One of the grandest works of ancient art which have come down to us, is the colossal statue of the Farnese Hercules. The hero of ancient mythology is represented in an erect form, leaning on his club, and ready for action; but at the moment every one of the well-developed muscles of his ponderous frame is fully relaxed, and the figure is a perfect ideal of repose, yet a wonderful embodiment of power. Here in this antique we have most perfectly typified the passive condition of oxygen, the hero of the chemical elements. Raise now the temperature to a red heat, and in a moment all is changed. The dormant energies of its mighty powers are aroused, and it rushes into combination with all combustible matter, surrounded by those glorious manifestations of light and heat which every conflagration presents." — COOKE.

130. *Combustion is Self-sustaining.* — It is not necessary to arouse any large amount of oxygen to activity in order to insure the continuance of the combustion. When, for instance, a lighted match is held to the wick of a candle, it excites but a few molecules of oxygen to activity. These few rush into combination with the elements of the candle, and by so doing develop sufficient heat to awaken the activity of more oxygen, which in

turn enters into combination and develops more heat. In this way a supply of active oxygen is maintained until the candle is consumed.

The slowness of combustion depends upon the fact, that the oxygen is mixed with the inert nitrogen in such proportions as generally to "restrain the awakened energies of the fire-element within the narrow limits which man appoints."

131. *The Point of Ignition.* — Different substances begin to burn at very different temperatures. This is well illustrated in the kindling of a coal-fire. Shavings are put into the grate first, then kindling-wood, then charcoal, and finally hard coal. The shavings are lighted by means of a match. The match is a bit of dry, soft wood, one end of which is covered with sulphur and tipped with phosphorus. It is a well-known fact, that when two bodies are rubbed together, heat is developed. On striking the match, sufficient heat is developed by the friction to ignite the phosphorus, which takes fire at a temperature of about 150° Fahrenheit. The phosphorus in burning develops heat enough to ignite the sulphur, which burns at a temperature of about 500°. The burning sulphur develops heat enough to ignite the wood of the match; the match, to ignite the shavings; the shavings, the kindling-wood; the kindling-wood, the charcoal; and the charcoal, the hard coal, which requires the temperature of a full white heat to set it on fire.

132. *The Products of Combustion are not always Gaseous.* — In the burning of coal-gas and of a candle, the products are wholly gaseous, and in the burning of wood they are mainly gaseous; but when metals, as copper and iron, burn in oxygen, the products of their combustion are *solid.*

133. *Magnesium and some other Metals will burn in the Air.* — A magnesium wire will burn brightly in the

air (96). Calcium, also, burns readily in the air (113), as aluminium (97) does, if pulverized and strongly heated.

It will be noticed that all these elements are rare in their free state. Many of the rare metals, as potassium, sodium, magnesium, and calcium, are as combustible as carbon; but they differ from carbon in giving rise to solid products while burning.

134. *Oxygen is not the only Supporter of Combustion.*—If a piece of gold-leaf or Dutch foil be dropped into a jar of chlorine, it vanishes in a flash of light (53). Here the burning consists in a combination with chlorine. Tin and copper foil and pulverized antimony will also burn in chlorine. Many metals will burn in the vapor of sulphur; the burning being then a combination of the metal with sulphur.

135. *The Materials of the Earth's Crust are chiefly Chemical Compounds.*—We have seen that when marble is acted upon by muriatic acid (42) we obtain *carbonic acid*, CO_2, and *calcic chloride*, CaCl. The calcium, carbon, and oxygen of these compounds must come from the marble, or *calcic carbonate*, $CaCO_3$.

We find, then, in marble two very combustible substances, calcium and carbon, combined with oxygen. Limestone has the same composition. Calcium, then, which is a very great rarity in a free state, is a very abundant element in nature.

It has been found that the rocks and solid matter of the earth are made up chiefly of such combustible elements as potassium, calcium, magnesium, aluminium, and carbon, combined with oxygen. Many of the rarest and most costly metals, then, are the most abundant in nature. The fact that they are so rare in a free state is due to their extreme combustibility. It is very difficult to separate them from oxygen, and no less difficult to keep them separate.

136. *The Present Materials of the Earth are Products of Combustion.* — We see, then, that water, the most abundant of liquids, is made up of the very combustible elements, hydrogen and oxygen; while the solid materials of the earth are composed chiefly of the very combustible elements, potassium, magnesium, calcium, aluminium, carbon, and silicon, in combination with oxygen. These materials are the products of combustion. There must have been a time when these elements existed together in a free state. Then, by some means unknown to us, the mass took fire; and the conflagration raged until all the materials were consumed. There was more oxygen than was needed for the combustion; and this is now found in the air in a free state. Oxygen was the most abundant of all the elements, since it alone makes up half the weight of the solid earth, eight-ninths the weight of the water, and one-fifth the weight of the atmosphere.

The diagram below (from Cooke) illustrates what has been said of the composition of the earth. It will be seen, that, of the 65 elements, 13 alone make up at least .99 of its known mass, while the other 52 do not constitute altogether more than .01.

<table>
<tr><td rowspan="4">Oxygen.
½</td><td colspan="2">Silicon.
¼</td></tr>
<tr><td colspan="2">Aluminium.
Magnesium.
Calcium.</td></tr>
<tr><td colspan="2">K Na Fe C</td></tr>
<tr><td>S H Cl N</td><td>52 others.</td></tr>
</table>

SLOW COMBUSTION.

137. *Partially Active Condition of Oxygen.* — We have seen (11, 12) that, besides its active and passive states, oxygen exists in a third, or intermediate, condition, in which it is *partially* active. In this form it plays a very important part in nature. It acts silently and slowly, taking months, and even years, to accomplish its work; but the results "far surpass in true grandeur those dazzling displays of power which the fire-element manifests when fully aroused."

138. *Decay.* — When wood, or any other *organic* substance (that is, any compound of *vegetable* or *animal* origin) is exposed to moist air, it *decays.* It first becomes rotten, and then slowly disappears. This decay consists in a gradual union of the substance with oxygen; in other words, a *slow combustion.* A log of wood which rots in the forest undergoes the very same change as one which is burnt on the hearth; the sole difference being, that, while the latter burns up in a few hours, the former is consumed only after the lapse of many years.

Wood, like other vegetable compounds, consists mainly of carbon, hydrogen, and oxygen. When it burns on the hearth, its carbon and hydrogen combine with the oxygen of the air, forming carbonic acid and water, which pass off in the smoke. The hydrogen is more combustible than the carbon, and therefore burns first, leaving the carbon in the form of glowing coals. These are consumed in their turn, gradually smouldering away, until nothing is left but a little ashes.

Quite the same process is going on with the decaying log in the forest. In decay, as in burning, the hydrogen of the wood unites with the atmospheric oxygen sooner than the carbon does; hence, in this stage of its decay, the wood becomes darker, and more like charcoal. At length, both the hydrogen and the carbon are burnt; and the log is slowly converted into carbonic acid and water, leaving only a handful of earth as the *ashes* of this lingering combustion. Even the *heat* generated in this form of burning is found to be precisely the same as in ordinary combustion; the only difference being that, in the latter case, it is all set free in a few hours, while in the former it is so slowly developed that it escapes our notice.

139. *Causes of Decay.* — Green wood decays much sooner than dry wood. Indeed, if wood be kept perfectly dry, it will not decay for ages. In the dry climate of Egypt, wooden mummy-cases have been preserved for more than three thousand years.

The decay of the green wood is due to the presence of what are called *albuminous* substances. The most important of these is *vegetable albumen*, and it is essentially the same thing as the albumen (or *white*) of an egg. We have said that most vegetable substances are made up of carbon, hydrogen, and oxygen; but these albuminous compounds, which form only a small part of the bulk of plants, contain an additional element, *nitrogen*. We have seen that the compounds of nitrogen are generally very unstable, and these albuminous substances are peculiarly so. In the presence of moisture they soon *putrefy*, or break up into simpler compounds. The oxygen of the air takes no part in this process, but, by contact with the putrefying substance, it is, in some mysterious way, awakened to the state of partial activity mentioned above. It is thus enabled to attack and consume the wood and all the other organic compounds present. The decay, or slow combustion, once begun, is self-sustaining; fresh portions of oxygen being continually roused to activity by the process itself, precisely as in ordinary burning (130).

When the nitrogenized vegetable compounds are *burnt*, the nitrogen passes off in a *free* state; when they *decay*, combined with hydrogen as *ammonia*.

140. *Rusting.*—The slow combustion of metals is called *rusting*, and the oxide formed is called *rust*. All the familiar metals, except silver, gold, and platinum, are tarnished on exposure to the air; that is, they become covered with a film of rust, or oxide.

That *heat* is developed by rusting, as by other kinds of slow combustion, is shown by the fact that if a large pile of iron-filings be moistened and exposed to the action of the air so that they rust rapidly, the temperature rises perceptibly.

A remarkable case of heat developed by rusting occurred in England during the manufacture of a submarine electric cable. The copper wire of the cable was covered with gutta-percha, tar, and hemp, and the whole enclosed in a casing of iron wire. The cable, as it was finished, was coiled in tanks filled with water: these tanks leaked, and the water was therefore drawn off, leaving about 163 nautical miles of cable coiled in a mass 30 feet in diameter (with a space in the centre 6 feet in diameter) and 8 feet high. It rusted so rapidly that the temperature in the centre of the coil rose in four days from 66° to 79°, though the temperature of the air did not rise above 66° during the period, and was as low as 59° part of the time. The mass would have become even hotter, had it not been cooled by pouring on water.

141. *Spontaneous Combustion.*—When charcoal which has been finely pulverized for making gunpowder is exposed in large heaps, the oxygen of the air combines with it slowly at first; but, as the heat developed accumulates, the oxidation becomes more rapid, until in some cases the mass takes fire and burns.

So too, when cotton or tow, which has been used for

wiping machinery, and has become saturated with oil, is laid aside in heaps, it begins to oxidize slowly; but the heat developed makes the combustion more and more rapid, until sometimes the heap bursts into a flame.

This rapid combustion, developed gradually from slow combustion, is called *spontaneous combustion*.

142. *Respiration.* — The albuminous or *nitrogenized* compounds, which form but a small part of the plant, make up almost the entire bulk of the animal, so that animal substances are even more prone to decay than vegetable. And this decay is not confined, as we might suppose, to *dead* animal matter, but is constantly going on in the *living* animal. The only difference is, that in the latter case the loss from decay is continually repaired, while in the former case it is not. The materials for the repair of the living body are furnished by the *food*.

This food is mainly made up of three classes of substances: (1) *non-nitrogenized*, as starch and sugar; (2) *nitrogenized*, as lean meat; and (3) *fatty* substances, as butter. All three kinds are absolutely necessary to the life of man. They are all contained in *milk*, which may be regarded as "the type of animal food." It contains *sugar*, which belongs to the first of the above classes; *caseine*, or *curd*, which belongs to the second; and *butter*, which belongs to the third. Bread also contains all three kinds of food; being made up of (1) *starch*, (2) *gluten* (the most important nitrogenized constituent of wheat and other grain), and (3) a small quantity of *oil*. A man cannot live for any length of time on any one kind of food, as starch or butter, or on any mixture of kinds of food, which does not contain all three classes of substances.

The different classes of food serve different purposes in the body. The *nitrogenized* and a part of the *fatty* substances supply the waste which results from the action of the varied machinery of the body. They renew the muscles, sinews, and nerves, which are constantly wearing out. On the other hand, the non-nitrogenized substances, as starch and sugar, probably take no part in repairing the machinery, but merely furnish fuel to keep up the heat of the body. To produce heat they must be burnt up, and we shall see that the process is another example of *slow combustion.*

Starch and sugar make up by far the greatest part of this fuel, and in fact of our food generally. They are almost identical in composition, and starch can readily be converted into sugar. When taken into the stomach, starch undergoes this change, and the sugar then readily dissolves in the water present. The solution is absorbed by the veins, and becomes mingled with the blood, which carries it to the heart. By the heart, which is made up of two force-pumps, the blood is forced through the lungs. These are composed of millions of little membranous bags or *air-cells*, closely packed together, and all connected by means of tubes with the windpipe, and thus with the nose and mouth. The membrane of the cells is very thin, so that they are easily compressed. The whole mass of the lungs is also very elastic, and by the action of muscles they are alternately expanded and contracted as we breathe. When they expand, the air from without rushes in and fills the cells; and when they contract, this air is forced out again.

We have said that the blood charged with sugar is forced through the lungs. The tube, or *artery*, which conveys it, divides and subdivides, until it is reduced to very small capillary tubes, which form a delicate network on the surfaces of the air-cells. The walls of these capillaries are very thin, and the oxygen of the air readily passes through them, by a process called *osmose*, and mingles with the blood. At the same time, and in the same way, carbonic acid is given out by the blood, and mixes with the air. The blood, holding in solution both sugar and oxygen, now goes back to the heart; and by the second force-pump it is sent to all parts of the body. In the mean time, the sugar is burnt up by the oxygen which was absorbed in the lungs.

"Sugar, like wood, consists of carbon, hydrogen, and oxygen. The last two are present in the proportions to form water, so that sugar may be said to be composed of charcoal and water. Of these two substances the charcoal only is combustible. This, during the circulation of the blood, is slowly burnt up by the dissolved oxygen, and converted into carbonic acid, which remains in solution until it is discharged, when the blood returns again to the lungs, or else escapes through the skin." *

Respiration, then, is a kind of combustion, in which "the fuel is sugar, and the smoke carbonic acid and aqueous vapor." The presence of carbonic acid in the air, from the lungs, may be proved by breathing through a glass tube into lime-water, which soon becomes milky. The presence of the watery vapor may be shown by breathing upon any cold substance.

The weight of carbonic acid breathed out by a full-grown man, in a day, varies from 1 to 3 pounds, or from 9 to 27 cubic feet; and the weight of carbon burnt is from 5 to 15 ounces. The amount of heat produced is of course the same as would be set free by burning the same weight of charcoal in a stove. The temperature of the body is thus kept above that of the air; the heat

* Cooke's "Religion and Chemistry." The greater part of §§ 135–137 and 140–142 has been condensed from the 3d and 4th Lectures in this admirable series. We cannot too strongly commend the book to teachers. They will be certain to draw from it, for the purposes of oral instruction, far more largely than our limits have permitted here.

of the blood, even in the coldest climate, being maintained at 96°.

"In regulating the temperature of his body, man follows instinctively the same rules of common sense which he applies in warming his dwellings. In proportion as the climate is cold, he supplies the loss of heat by burning more fuel in his lungs, and hence the statements of arctic voyagers, who have told us that twelve pounds of tallow-candles make only an average meal for an Esquimaux, are not inconsistent with the deductions of science."

143. *Slow and Rapid Combustion of Sugar.*—We may compare the rapid combustion of sugar in air with its slow combustion in the body, by the following experiment: Take 2 ounces of pulverized sugar, or the average quantity burnt in the body of a man in an hour, and mix it with 5½ ounces of pulverized potassic chlorate. We have then the sugar and the solidified oxygen of the chlorate mingled as in the blood; but the oxygen is passive, and will not combine with the sugar, until roused to activity. A single drop of sulphuric acid let fall upon the mixture serves to awaken its dormant energy; and the mass is consumed in an instant, with intense evolution of heat and light. The amount of heat concentrated in this momentary burst of flame is no greater than would have been generated in the blood in the course of an hour.

"The splendid displays of combustion arrest our attention by their very brilliancy, while we overlook the silent yet ceaseless processes of respiration and decay, before which, in importance and magnitude, the greatest conflagrations sink into insignificance. These are but the spasmodic efforts of nature; those, the appointed means by which the harmony and order of creation are preserved."

144. *The Daily Consumption of Oxygen in Nature.* — "Faraday has roughly estimated that the amount of oxygen required daily, to supply the lungs of the human race, is at least one thousand millions of pounds; that required for the respiration of the lower animals is at least twice as much as this, while the always active processes of decay require certainly no less than four thousand millions of pounds more, making a total aggregate of seven thousand millions of pounds required to carry on these processes of nature alone. Compared with this, the one thousand millions of pounds which, as Faraday estimates, are sufficient to sustain all the artificial fires lighted by man, from the camp-fire of the savage to the roaring blaze of the blast-furnace, or the raging flames of a grand conflagration, seem small indeed.

Amount of Oxygen required Daily.

Whole population	1,000,000,000
Animals	2,000,000,000
Combustion and fermentation .	1,000,000,000
Decay and other processes . .	4,000,000,000
Oxygen required daily . . =	8,000,000,000 lbs.

"How utterly inconceivable are these numbers, which measure the magnitude of nature's processes, — eight thousand millions of pounds of oxygen consumed in a single day! When reduced to tons, the number is equally beyond our grasp; for it corresponds to no less than 3,571,428 tons. If such be the daily requisition of this gas, will not the oxygen of the atmosphere be in time exhausted? It is not difficult to calculate approximately the whole amount of oxygen in the atmosphere. It is equal to about 1,178,158 thousand millions of tons; a supply which, at the present rate of consumption, would last about nine hundred thousand years."

THE GROWTH OF PLANTS.

145. *The Embryo Plant in the Seed.*—How do plants grow? How does the tiny seed become the leafy herb? How does the little acorn develop into the giant oak? We shall get the best answer to these questions by tracing a plant through its whole growth.

The seed is formed in the flower, the essential parts of which are the *stamens* and the *pistil.* The *stamens* bear the *anthers*, and these contain the *pollen.* The pistil is in the centre of the flower. It encloses in its *ovary* the *ovules*, which, when ripened, become *seeds.*

In the ripe seed we find the *embryo*, which is a miniature plant with stem and leaves. How has this embryo been formed?

At a certain time a little cavity is formed in the centre of the ovule within the pistil. This cavity is called the *embryo sac*, and is marked *s* in Figure 20. Within this sac, at its upper end, we see a minute body or *vesicle*, *v.* This is the first germ of the embryo; but in order that it may begin its development, it must be acted upon by the pollen. This we see in the form of small grains, *a*, resting on the top of the pistil. It has fallen from the anthers, and lodged here; and now it sends out a very fine and delicate tube, the *pollen-tube*, which pierces the tissue of the pistil, and extends itself until it reaches the vesicle, *v.* Its contact with this microscopic particle of matter has the mysterious power of making it begin to *grow*, and thus form the embryo.

Fig. 20.

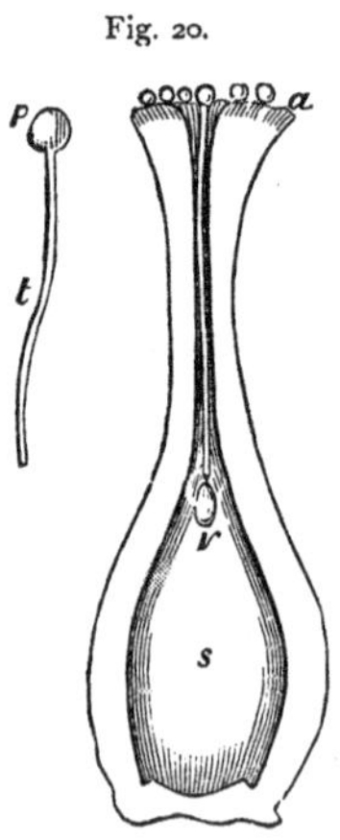

The vesicle is at first a single *cell;* that is, a membranous globule, filled with liquid, in which minute particles can sometimes be discerned. After it has been acted upon by the pollen-tube, it enlarges somewhat, and a partition forms across its interior, dividing it into two cells. Each of these grows and divides in like manner, and thus a cluster of cells is formed. After a time the mass begins to take a definite shape. One end becomes the *radicle*, or the beginning of the *root* of the plant; and the other divides into two parts, which develop into the *cotyledons*, or *seed-leaves*. It is now a perfect embryo, a miniature plant, with root, stem, and leaves, but still shut up in the seed.

146. *The Plantlet.* — The growth of the plant in the seed, and out of the seed, is essentially the same process. The same division and multiplication of cells continues, and all the parts of the plant are formed by the clustering or *aggregation* of these cells. The cells are too minute to be distinguished with the naked eye, but the microscope shows them in every portion of the plant. Figure 21 shows a thin slice of a rootlet, cut crosswise and lengthwise, as it appears when highly magnified. We see that the whole structure is *cellular;* that is, made up of cells crowded together. A single cell is represented in Figure 22. The natural shape of the cell is spherical; but when cells are clustered or crowded together, they compress one another into the shape in which we see them in these figures. Where they are not thus squeezed, they are generally found to be spheres, more or less perfect. They vary in size, from about the thirtieth to the thousandth of an inch in diameter.

Fig. 21.

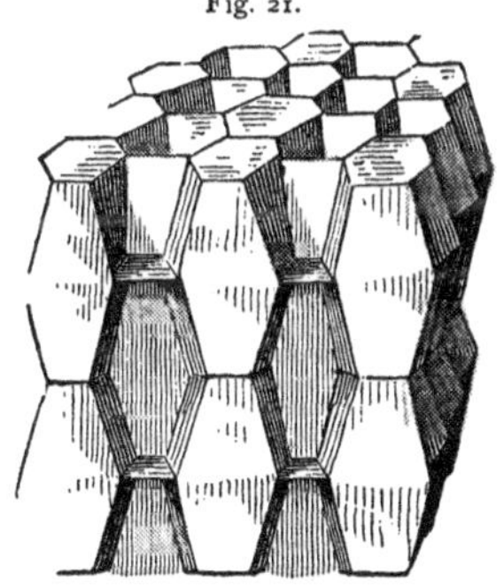

At first the cells are all alike in shape and in texture; but they are greatly modified in both respects in forming the varied tissues of the plant. They may be lengthened out into tubes, as in the fibres of cotton, which are single cells thus drawn out. The soft, thin membrane which encloses them may become hard and thick, as in the shell of a walnut or the tough wood of an oak. They may be closely crowded together, and nearly filled up with solid matter; or they may be loosely interlaced, and form vessels to contain the vegetable juices. Their contents are as varied as their structure, and, seen through the transparent walls, give rise to all the manifold colors of leaf and flower and fruit.

Fig 22.

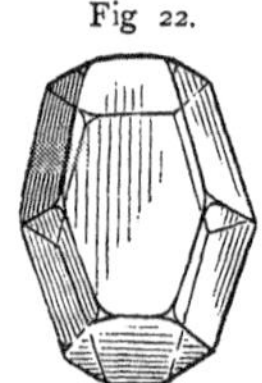

Figure 23 (from Wood's "Class-book of Botany") shows a few of the varied forms which cells assume. At the top we have wood-cells from the fibre of flax. Below

Fig. 23.

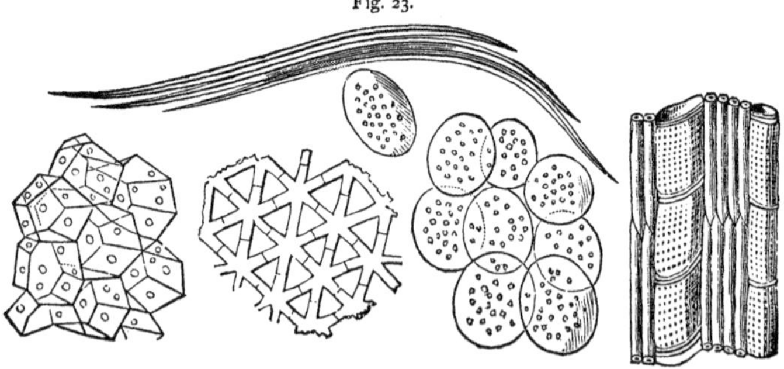

are represented the many-sided cells of the pith of elder; *stellate* (star-shaped) cells of the pith of the rush; spherical cells of the houseleek; and wood-cells of the oak.

147. *Organic Structure.*—This cellular structure is the characteristic of *organic* beings; that is, *plants* and

animals, as distinguished from *inorganic* things, or *minerals;* and hence it is called *organic* structure. The cell is "the simplest form of organic life," whether vegetable or animal. Cells are the *units* of which the most complicated organic tissues are made up. They are essentially different, then, from the particles into which an inorganic or mineral substance may be divided. Each of these particles has all the properties of the mass from which it was separated; and if we divide it into yet smaller particles, the same will be true of those; and so on indefinitely. But if we cut a cell in two, the halves do not have the properties of the whole; they are not smaller units, but the fragments of a unit.

148. *The Food of Plants.*—Plants get their food from the earth and the air, but much the greater part from the air. We know that there are plants which flourish in the most barren soil, or even upon the naked rock, and that some live and grow suspended in the air, and having no contact with the earth. Even those which demand a rich soil are indirectly indebted to the air for what they draw from the earth. The soil owes its fertility to the decomposition of organic matter; and this organic matter was originally produced by plants which had no rich soil to draw from, but were dependent mainly upon the air.

The atmosphere, then, is the storehouse from which plants directly or indirectly obtain nearly all their food. The portion which is purely of earthy origin is always insignificant, and often it is nothing at all. In fact, plants give to the earth far more than they get from it. This is illustrated by the accumulation of vegetable organic matter in the soil wherever vegetation is undisturbed from year to year. In uncultivated fields and in primeval forests we often find a great depth of rich mould. The more rank and luxuriant the vegetation, the more rapidly this deposit increases, showing that the plants not only restore to the soil all that they have drawn from it, but are continually transferring fresh matter from the aerial storehouses to the earth.

But while the soil is enriched by undisturbed vegetation, it is impoverished by agriculture. The farmer carries away the crop from the field, with all that it has taken from both the earth and the air. The land cannot yield in this way year after year, unless he follows the example of nature, and restores to the soil an equivalent for what he removes. This he can do by the use of *manure*.

149. *The Earthy Portion of the Plant.*—If we burn wood or any other vegetable substance, almost all of it is dissipated into air. But a little ashes will remain; and these represent the earthy or *inorganic* portion of the plant. They consist mainly of alkaline chlorides, potash, soda, silica, metallic phosphates, calcic and magnesic carbonates, and ferric and manganic oxides. These are dissolved in the water which soaks through the soil and which is taken up by the roots of the plant. Much of the water is evaporated through the leaves, but the substances which it held in solution remain behind, and thus gradually accumulate in the tissues of the plant.

Since the plant must obtain from the soil the inorganic materials it needs, it is evident that it will flourish only in a soil containing those materials. This explains why certain plants thrive only in certain situations. A locality may be fertile for some species of vegetation and barren for others. The pines, which need little alkaline matter, will flourish in a sandy soil containing little alkali; but the maples and elms, which require a good deal of potash, cannot live in such a soil.

We have said that the farmer, who carries away the produce of the field with all that it has drawn from the soil, must restore in the form of manure an equivalent for what he removes, or the field will soon become impoverished.

"A medium crop of wheat takes from one acre of ground about 12 pounds; a crop of beans, about 20 pounds; and a crop of beets, about 11 pounds of phosphoric acid, besides a very large quantity of potash and soda. It is obvious that such a process tends continually to exhaust arable land of the mineral substances useful to vegetation which it contains, and that a time must come, when, without supplies of such mineral matters, the land would become unproductive from their abstraction. . . . In the neighborhood of large and populous towns, for instance, where the interest of the farmer and market-gardener is to send the largest possible quantity of produce to market, consuming the least possible quantity on the spot, the want of saline principles in the soil would very soon be felt, were it not that for every wagon-load of greens and carrots, fruit and potatoes, corn and straw, that finds its way into the city, a wagon-load of dung, containing each and every one of these principles locked up in the several crops, is returned to the land, and proves enough, and often more than enough, to replace all that has been carried away from it." — BOUSSINGAULT.

This renewal of fertility is sometimes attained by letting the field lie *fallow*, or uncultivated, for one or more years. The inorganic materials of the soil are mainly furnished by the gradual disintegration of the *rocks;* and while the field lies fallow this process is going on, under the influence of the oxygen and carbonic acid of the air, aided by the rains and changes of temperature. In this way, fresh portions of the rocks or of their ruins are rendered soluble and thus fitted for the nourishment of plants.

An *alternation of crops* may answer the same purpose as letting the field lie fallow. For instance, wheat and potatoes may be raised on the ground in alternate years. The wheat requires a large amount of silica and alkaline matter, while the potatoes take up no silica. The renewal of the soluble silica in the soil therefore goes on, while the potatoes are growing, as it would if the field were fallow.

Some soils abound in silicates so readily decomposed, that in every one or two years a sufficient supply for a crop of wheat becomes soluble. In Hungary there are large districts where wheat and tobacco have been raised alternately upon the same soil for centuries, the land never receiving back any of the mineral matter which is carried away with the crops. On the other hand, there are fields in which the amount of soluble silica required for a single crop of wheat is not separated from the insoluble masses in the soil in less than three or four years.

150. *The Organic Portion of the Plant.* — When we burnt the vegetable matter, much the greater part of it passed off into the air. This was the *organic* portion of the plant. It was taken from the air, and the burning has given it back in the very form in which the plant found it there.

We have learned that all the vegetable tissues, however varied in their texture and consistency, are almost entirely made up of the same kind of cells. The substance of which these cells are made is called *cellulose*, or *woody fibre;* and it is a compound of carbon, hydrogen, and oxygen, $C_{12}H_{20}O_{10}$. We have also learned that the plant contains small quantities of *albuminous* compounds (139), which, in addition to the elements just named, contain *nitrogen*. These four elements, then, are essential to the growth of the plant, and must be contained in its food.

151. *The Plant gets Hydrogen and Oxygen from Water.* — The hydrogen and the oxygen of the cellulose are obtained from the *water* which the plant takes in, not only through its roots, but through its leaves. It will be noticed that water contains hydrogen and oxygen in the same proportions as cellulose does.

152. *The Plant gets Carbon from Carbonic Acid.* — We have learned that carbon is a solid, and insoluble in water. Even if it were reduced to the finest powder and mixed with the water, it could not be taken up by the plant, since nothing but liquids and gases can pass through the walls of the cells. It must, then, be furnished to the plant in some liquid or gaseous compound, like the carbonic acid, which, as we have seen, is one of the gases mixed in the atmosphere.

Now if a leafy plant be placed under a glass vessel and set in the sunshine, and a stream of carbonic acid be made to pass slowly over it, it is found that a part of the carbonic acid is removed and replaced by oxygen. The plant absorbs the carbonic acid, decomposes it, retains the carbon, and exhales the oxygen.

We conclude, then, that it is from the carbonic acid in the atmosphere that plants get their carbon. In the leaves, under the influence of sunlight, the carbonic acid is decomposed, the carbon stored away in the plant, and the oxygen given back to the air.

Since carbonic acid is soluble in water, it is probable that plants also obtain a part of their carbon through the roots. Plants which live under water must get all their carbon from the gas dissolved in the water.

153. *The Plant gets Nitrogen from Ammonia.* — We have learned (123) that there is ammonia in the atmosphere, and (37) that this gas is very soluble in water. Hence it is washed out of the air by the rain, and, thus dissolved in water, is taken up by the roots of plants.

154. *The Growth of Plants is a Chemical Process carried on in the Leaf by the Sunlight.*—We have seen that the plant takes in water, carbonic acid, and ammonia; that it decomposes these substances; and that from their elements it elaborates all the organic compounds which enter into its own structure. The leaf is the laboratory in which this chemical process is conducted, and since it is only in sunshine or bright daylight that the work goes on, it is evident that the sunlight is the agency by which it is accomplished.

"The sun's rays, acting on the green parts of the leaf, give to them the power of absorbing water, carbonic acid, and ammonia, and of constructing from the materials thus obtained the woody fibre, starch, sugar, and other compounds of which the plant consists. We have analyzed the woody fibre, and we know that it is composed of charcoal and water. Nineteen ounces of wood contain nine ounces of charcoal and ten ounces of water. Moreover, the amount of charcoal required to make nineteen ounces of wood is contained in thirty-three ounces of carbonic acid. If, then, we add together thirty-three ounces of carbonic acid and ten ounces of water, and subtract from this sum twenty-four ounces of oxygen, we shall have just the composition of wood. This is what the sun's light accomplishes in the leaves of the plant. It decomposes the carbonic acid, and unites its carbon to the elements of water to form the wood."—COOKE.

What is here stated to be true of wood is equally true of other vegetable products. If in their production carbonic acid and water alone take part, we have such substances as woody fibre, starch, sugar, and gum; and these make up nine-tenths of all vegetable structures. If the ammonia is likewise employed in the process, we have *nitrogenized* products, like albumen and caseine.

The force which the sunbeam exerts in the decomposition of carbonic acid in the leaf is very remarkable, since it overcomes the intense affinity of oxygen for carbon.

"In order to decompose carbonic acid in our laboratories, we are obliged to resort to the most powerful chemical agents, and to conduct the process in vessels composed of the most resisting materials, under all the violent manifestations of light and heat, and we then succeed in liberating the carbon only by shutting up the oxygen in a still stronger prison; but under the quiet influences of the sunbeam, and in that most delicate of all structures, a vegetable cell, the chains which unite together the two elements fall off, and, while the solid carbon is retained to build up the organic structure, the oxygen is allowed to return to its home in the atmosphere. There is not in the whole range of chemistry a process more wonderful than this. We return to it again and again, with ever increasing wonder and admiration, amazed at the apparent inefficiency of the means, and the stupendous magnitude of the result. When standing before a grand conflagration, witnessing the display of mighty energies there in action, and seeing the elements rushing into combination with a force which no human agency can withstand, does it seem as if any power could undo that work of destruction, and rebuild those beams and rafters which are disappearing in the flames? Yet in a few years they will be rebuilt. This mighty force will be overcome; not, however, as we might expect, amidst the convulsion of nature, or the clashing of the elements, but silently, in a delicate leaf waving in the sunshine." — COOKE.

155. *Plants purify the Air for the Respiration of Animals.* — We have learned (144) that by the various forms of combustion going on in nature, eight thousand millions of pounds of oxygen are daily removed from the atmosphere; and that an immense amount of carbonic

acid is pouring back into the air as one of the chief products of this combustion. The atmosphere, then, is continually losing the element which is essential to the support of animal life, and receiving in its place a gas which, even when largely diluted with air, is a deadly poison to animals.

But we have also learned that during the daytime plants are constantly drinking in this carbonic acid, and giving out an equal bulk of pure oxygen. They are thus purifying the air for the respiration of animals. They breathe in what animals breathe out, and breathe out what animals breathe in; so that the air is kept in a fit state to sustain both forms of organic life.

An Oriental fable narrates that a nightingale and a rose-tree were once imprisoned in a cage of glass, and long lived together there, happy in each other's society. But at last the rose-tree withered and died; and the bird soon pined away, and perished with grief at the loss of its beloved companion. The fancy of the poet is in perfect harmony with the facts of science, and prettily illustrates the dependence of animals upon plants for the very breath of life.

156. *Plants prepare Food for Animals.* — We have seen that the plant draws its food directly from the earth and the air. It has the power of forming organic compounds out of inorganic matter. The animal, on the other hand, has no such power. It can elaborate its tissues only from organic matter, which it receives ready-made from the vegetable kingdom. It may get this matter either directly from the plant, as when it eats vegetable food, or, indirectly, as when it eats animal food; but in either case the origin is the same. By the process of digestion the organic compounds originally prepared by the plant are converted into bones, muscles, nerves, or whatever else enters into the struc-

ture of the animal. In this process they do not undergo any radical change, but merely "become parts of more finely organized tissues." We find in the blood albumen and caseine, having precisely the same composition as when prepared from potatoes, and the substance of the muscles does not differ essentially from the gluten of wheat-flour.

157. *The Relations of Plants and Animals to each other and to the Air.* — The plant, then, is a *producer* of organic materials; the animal, a *consumer* of these materials. "While the plant is a true apparatus of reduction, the animal is a true apparatus of combustion, in which the substances it derived from the vegetable are burnt, and restored to the atmosphere in the form of carbonic acid, water, and ammonia, ready to be again absorbed by the plant and to repass through the phases of organic life. Our bodies are furnaces, — furnaces continually burning, — whose fuel is our own flesh, and the smoke of whose fires is the food of the plant. . . .

"When the foundations of the globe were laid, there were collected in the atmosphere all the essential elements of organized beings. From this inexhaustible storehouse the plant absorbs water, carbonic acid, and ammonia, which were placed there for its use, and which have been made to serve as its nourishment and food. It is the special office of the plant to elaborate from these few mineral substances, and a small amount of earthy salts, all the materials of organized beings. The animal receives these crude materials already prepared, and builds with them its various tissues; but no sooner are the cell-walls finished, and the structure ready to discharge its vital functions, than it is consumed by almost the very act which gave it life. The carbonic acid, water, and ammonia are restored to the atmosphere, and the cycle is complete." — COOKE.

NATURAL VEGETABLE PRODUCTS.

158. *Starch.* — *Starch* has already been mentioned as one of the most important products of vegetable growth. Its chemical composition is precisely the same as that of woody fibre, $C_{12}H_{20}O_{10}$. It consists of round or oval granules, varying considerably in size in different vegetables. The largest are about $\frac{1}{260}$th, and the smallest less than $\frac{1}{3000}$th of an inch in diameter. In the potato (Figure 24) they are considerably larger than in wheat-

Fig. 24. Fig. 25.

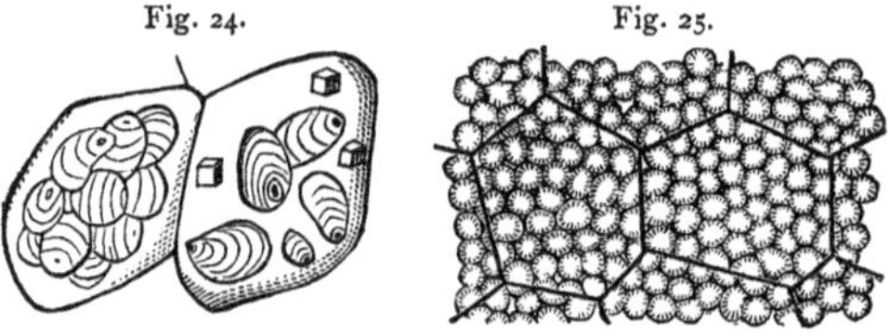

flour (Figure 25). These granules consist of concentric layers, formed one after another, about a *nucleus* or centre. Lines, indicating these layers, can sometimes be seen on the surface of the larger grains, as in those of potato starch in the figure. Since the grains from the same kind of plant are tolerably uniform in size and shape, while they vary much in different species, the microscope will show to what plant starch-grains belong. Adulterations of arrowroot, and of other starchy substances, may thus be detected.

Starch is heavier than water, and is insoluble in cold water, alcohol, or ether. If, however, it be placed in water at 150°, the grains swell up and burst, forming a paste or jelly. When long boiled in water, starch is converted into the soluble *dextrine* (162).

The test for starch is iodine, with which it forms a compound of a deep blue color (60). Bromine gives starch a brilliant orange tint.

Fig. 26.

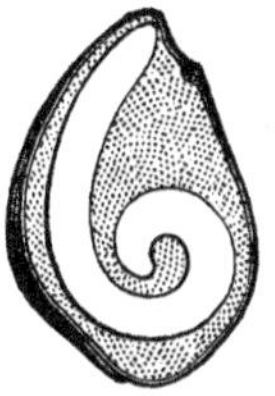

Starch is the form in which food is stored up in the plant for future use. In the seed, it furnishes the material for the growth of the embryo, before it can draw its nourishment from the earth and the air. It sometimes surrounds the embryo, as in the seed of the onion (Figure 26); and sometimes it is found in the *cotyledons*, or *seed-leaves*, of the embryo itself, as in the seed of the bean (Figure 27).

Fig. 27.

When the seed germinates, the starch is dissolved, and converted into dextrine, and this into sugar, which is even more soluble. It is then dissolved in the sap, which conveys it wherever it is needed for the growth of the infant plant.

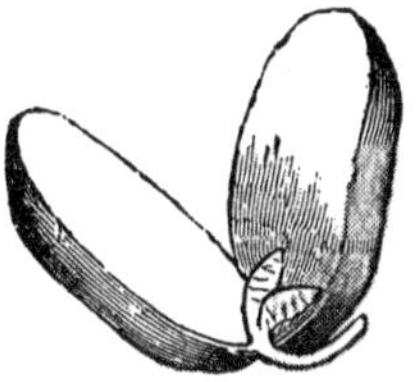

Several forms of starch are extensively used as articles of food. *Arrowroot* is obtained from the roots of a tropical plant, which is cultivated in the West Indies, especially in Bermuda and Jamaica, and to some extent in the East Indies and in Africa. *Tapioca* is also made from the roots of plants which grow in the West Indies, South America, and Africa. *Sago* is extracted from the pith of several species of palm-tree, in India and the islands of the Indian Archipelago. *Corn-starch*, *maizena*, and *farina*, and various other preparations of the kind, are made from the starch of Indian corn, wheat, etc. Potatoes yield from 12 to 27 per cent of starch; peas and beans from 33 to 36 per cent; wheat-flour, 56 to 72 per cent; rye-meal, 61 per cent; maize, 81 per cent; rice, 83 to 85 per cent.

Starch is also used in large quantities for laundry purposes; in the manufacture of cotton cloth, as a *sizing* for the thread; and in the making of dextrine and grape-sugar (161, 162).

159. *Sugar.* — Sugar, like starch, is a vegetable product, prepared for the future growth of the plant. It is found, dissolved, in varying proportions, in the juices of all plants. In ripe fruits it is quite abundant; pears containing 6 per cent; peaches, 16½ per cent; and cherries, 18 per cent. In rye-meal there is about 3½ per cent of sugar; in wheat-flour, from 4 to 8 per cent; in beet-root, from 5 to 8 per cent; and in dried figs, more than 60 per cent.

There are many kinds of sugar, of which the two most important are *cane-sugar* and *grape-sugar.*

160. *Cane-Sugar.* — Cane-sugar, $C_{12}H_{22}O_{11}$, is found in the juices of many plants. It is mostly prepared from the juice of the *sugar-cane;* but on the Continent of Europe, it is largely made from *beet-juice.* In tropical regions, it is also manufactured from the juice of the *date-palm;* and in the Northern United States it is obtained in considerable quantities from the sap of the *rock* or *sugar maple.*

In the manufacture of sugar from sugar-cane, the juice is first extracted from the canes by pressure, then mixed with a small quantity of slaked lime, and heated nearly to boiling. The lime serves to neutralize the acid in the juice, and also to remove the albuminous matter, which, if left in the sugar, would soon cause it to ferment and spoil. The juice is next evaporated in open pans to a thick syrup, which is allowed to cool and crystallize, and is then drained in perforated casks. The liquid drained off is known as *molasses.*

By this process, *raw* or *brown sugar* is obtained. This is *refined* by dissolving it in water, adding albumen (usually from ox-blood), and heating it. The albumen separates most of the impurities from the solution, which is then filtered through bone-black to remove the coloring matter. It is then boiled down in *vacuum pans*, large vessels from which the air is partially exhausted by pumps, in order that the evaporation may be carried on at a lower temperature than in open pans. By this process there is a saving of fuel, while the quantity of sugar obtained is increased. The concentrated syrup is crystallized in conical moulds, with an opening at the bottom for drainage, and thus becomes *loaf-sugar*.

When sugar crystallizes very slowly, it forms the large and very hard crystals seen in *rock-candy*. In making loaf-sugar, the syrup is frequently stirred, to prevent the formation of these large crystals.

If cane-sugar be heated to about 400°, it loses two atoms of water, and is converted into *caramel*, a dark brown substance, much used for coloring brandy and other spirits.

The manufacture of sugar from the sugar-cane, and other sources, is now one of the largest branches of human industry, but its importance is of comparatively modern date. Sugar was known to the Greeks and Romans only as a medicine or a curiosity. For several centuries after the Augustan age, we find scarcely any mention of it; and, even so late as the seventh century, Paul of Ægina describes it as "India salt, resembling common salt in color and consistency, but honey in taste and flavor." The sugar-cane appears to have been introduced into Europe by the Saracens, who cultivated it in Rhodes, Cyprus, Crete, and Sicily, in the ninth century. In the fifteenth century, it was transplanted into Madeira and the Canary Isles, whence it was probably brought to America. It became known in England in the fourteenth century; and, in 1329, it was sold in Scotland at one ounce of silver (equal to four dollars of our money) a pound. It did not become an article of ordinary consumption until the beginning of the seventeenth century. Since that time its use has very rapidly increased, and it seems destined to an indefinite extension. "It is so nutritious, wholesome, and agreeable, that there never can be a limit to its use, except in a prohibition or an inability to buy it. Men and nations differ widely in respect to most kinds of food, sauce, and drinks. Neither wheat, rice, flesh, nor potatoes can command unanimous favor. No article of housekeeping, save sugar, can be named, which is so universally acceptable to the infant and the aged, the civilized and the savage."

The annual production of sugar throughout the world is estimated at more than 5,000 millions of pounds, of which nearly 88 per cent is cane-sugar; about 7 per cent, beet-sugar; 4 per cent, palm-sugar; and about 1 per cent, maple-sugar. According to Chambers's Encyclopædia (1867), the yearly consumption *for each inhabitant*, in several great countries, is as follows: Russia, 1½ lbs.; Austria, 1½ lbs.; France, 4 lbs.; Belgium, 6 lbs.; Great Britain, 30 lbs.; United States, 40 lbs.

161. *Grape-Sugar.* — Grape-sugar, or *glycose* (less properly spelled *glucose*), $C_{12}H_{24}O_{12}$, is found in the juice of grapes, plums, cherries, figs, and many other fruits, and may often be seen in a crystalline form on raisins and dried figs. It likewise occurs in honey.

It is manufactured on a large scale, in Europe, from starch. A mixture of starch and water, at a temperature of about 130°, is made to flow gradually into a vat, containing boiling water, acidulated with one per cent of sulphuric acid. In about half an hour, the starch is converted into sugar. The sulphuric acid is neutralized with chalk, forming a deposit of calcic sulphate; and the clear solution of sugar is then boiled down and crystallized. Instead of starch, in this process, paper, flax, cotton and linen rags, sawdust, or any other form of woody fibre, may be used. The woody fibre is first converted into starch, and then into sugar.

Glycose is largely used in Europe for confectionery, for adulterating cane-sugar, and for the manufacture of beer and spirits.

162. *Dextrine.*—Dextrine, or *British gum*, $C_{12}H_{20}O_{10}$, is obtained by heating starch to about 400°; or by subjecting the starch to the long-continued action of dilute acids, at a high temperature. It is often used instead of gum-arabic in calico-printing, and for stiffening certain goods. It is also applied to the back of postage-stamps and other adhesive labels.

When wheaten bread is baked, a small quantity of starch is converted into dextrine on the outside of the crust, forming a brownish glazing.

163. *Cellulose and Gun-Cotton.*—*Cellulose*, or *woody fibre*, $C_{12}H_{20}O_{10}$, has already been described (150). It is insoluble in water, alcohol, or ether, but dissolves in an ammoniacal solution of cupric oxide (oxide of copper).

Gun-cotton, or *nitro-cellulose*, is made by the action of strong nitric acid upon cellulose. Cotton wool is immersed, in small portions at a time, in a mixture of equal volumes of strong nitric and sulphuric acids. It does not undergo any apparent change; but, on being washed and dried, it is found to be very inflammable. Six atoms of its hydrogen have been replaced by NO_2, so that its composition is now represented by $C_{12}H_{14}(NO_2)_6O_{10}$.

The use of gun-cotton, as a substitute for gunpowder, has been proposed, for the following reasons:—

(1) The explosive force of gun-cotton is, weight for weight, greater than that of gunpowder. (2) The products of the combustion of gun-cotton, being chiefly carbonic acid and nitrogen, are not so apt to foul the gun. (3) When moistened, it becomes inexplosive, and only requires drying to render it again explosive.

The reasons which render the general adoption of this substance doubtful are,—(1) its liability to explode on percussion; (2) the possibility of its spontaneous decomposition when kept for a length of time.

A variety of gun-cotton, containing a smaller proportion of NO_2, dissolves readily in a mixture of ether and alcohol, and yields a solution termed *collodion*. This is largely used for the purpose of forming a thin coating on glass to receive silver salts, upon which the photographic image is formed.

164. *Vegetable Parchment.*—When paper is immersed for a short time in a mixture of two volumes of oil of vitriol and one of water, and thoroughly cleansed by repeated washings with water, and finally with ammonia, it is changed into a substance resembling parchment, and often called *vegetable parchment.*

"The alteration which takes place in the paper is of a very remarkable kind. No chemical change is effected, nor is the weight increased; but it appears that a molecular change takes place, and the material is placed in a transition state between the cellulose of woody fibre and dextrine."

Vegetable parchment is now extensively used instead of parchment for legal and other documents. In some respects it is preferable to the old kind, for insects attack it less.

165. *Vegetable Acids.*—The acids of vegetable origin are almost innumerable. We shall describe here only a few of the most important.

Tartaric acid, $C_4H_6O_6$, is found, either free or combined with potash or lime, in tamarinds, grapes, pineapples, and many other vegetables. It is usually seen in the form of colorless crystals, which have an agreeable acid taste, and are soluble in water and alcohol. It is used in large quantities in calico printing and dyeing, and also in medicine, especially for making effervescent draughts. Some of its more important compounds are the familiar substances, *Rochelle salts*, *cream of tartar*, and *tartar-emetic*.

Citric acid, $C_6H_8O_7$, is usually prepared from lemons. It is also contained in gooseberries, currants, raspberries, strawberries, and other fruits. It is a crystalline solid, readily soluble in water, and has an intensely sour taste. It is used in silk-dyeing, in calico-printing, and in medicine.

Malic acid, $C_4H_6O_5$, is one of the most abundant of these vegetable acids. It is found in apples, from which (Latin, *malum*) it takes its name; in currants, barberries, rhubarb, and many other fruits and plants. It forms needle-shaped crystals, which are very soluble in water and alcohol. It is not used in the arts or in medicine.

Oxalic anhydride, C_2O_3, is not known in a free state. It is found in the juice of wood-sorrel (*oxalis acetosella*, from which it takes its name), and of many other plants, in combination with potash or lime. It forms a compound with water, $H_2C_2O_4$, which is *oxalic acid*. This is a crystalline substance, and is very poisonous. It is much used in cotton-printing and straw-bleaching, and is prepared in very large quantities by the action of caustic potash on sawdust. Crude potassic oxalate (oxalate of potash) is thus formed, which, by means of lime, is converted into the insoluble calcic oxalate (oxalate of lime), and this is decomposed by sulphuric acid.

Tannic acid and *tannin* are the names given by chemists to a number of compounds of C, H, and O, which are characterized by a well-marked *astringent* taste. They are soluble in water and alcohol, and form precipitates with most metallic oxides. Ferric salts yield black, or nearly black, precipitates, which are the basis of the ordinary writing *inks*. Tannic acid also precipitates *gelatine* (or *glue*); and it is by a similar process that it converts the gelatinous tissue of raw hides into *tanned* leather. The bark and leaves of most forest trees, such as the oak, elm, willow, horse-chestnut, and pine, and of many fruit-trees, as the pear and plum, contain considerable quantities of tannin. Coffee and tea likewise contain modifications of this compound; but the tannin in coffee, unlike that from all the other plants we have named, does not form the black salt with iron. A drop of tea on a knife-blade becomes inky at once, but a drop of coffee does not.

166. *Natural Fats and Oils.* — The natural oils and fats (fats being merely solid or semi-solid oils) are all compounds of organic acids, called the *fatty acids*, with a base called *glycerine* (167), and they are found in both plants and animals. *Cocoa-nut oil*, *palm oil*, and *nutmeg butter* are examples of vegetable fats.

The vegetable *oils* are divided into the *fixed oils*, which cannot be distilled without decomposition, and the *volatile oils*, which bear distillation without change. The *fixed* oils are divided into *drying* and *non-drying oils*. The former become dry and solid from oxidation, when spread out thin and exposed to the air; while the latter remain unaltered. The chief drying oils are those of linseed, hemp, poppy, and walnut, all much used in paints and varnishes; and the most important non-drying oils are olive oil, almond oil, and colza oil, which are extensively used in making soap, candles, and illuminating oils, in wool-dressing, and for many other purposes. Castor oil seems to form a connecting link between these two classes of oils, as it gradually becomes hard by long exposure to the air.

The *volatile* or *essential* oils are believed to constitute the odorous principles of plants. Most of them are nearly colorless when fresh, but darken on exposure to light and air. When long exposed to the air, they absorb oxygen, and become *resins*. They are largely used in the manufacture of perfumery; for flavoring confectionery and liquors, and for various other purposes in the arts; and in medicine.

167. *Soap.* — When the oils or fats are boiled with an alkali, they undergo the remarkable change called *saponification.* The fat is decomposed into a *fatty acid* and *glycerine* (166). The acid combines with the alkali to form *soap*, and the glycerine passes into solution. If the alkali be *soda*, a *hard* soap is formed; if *potash*, a *soft* soap. The soaps, then, like the fats, are true *salts.*

Fats may also be separated into acid and glycerine by distillation with steam, at a temperature between 500° and 600°.

Glycerine ($C_6H_{16}O_6$), is a colorless viscid liquid, of a sweet taste (whence its name, which is from a Greek word, meaning *sweet*), soluble in water and alcohol, but nearly insoluble in ether. It is used in medicine, in making copying-ink, and as a lubricating agent.

The most important of the fatty acids are *oleic*, *stearic*, and *palmic*, or *palmitic.* They are used on a large scale in the manufacture of candles.

168. *Wax.* — The various forms of vegetable wax resemble in many respects the fixed oils, but are quite different from them in chemical composition. They are solid or semi-solid substances; easily broken when cold, but soft and pliable when moderately warm, and melting below 212°. They are insoluble in water and cold alcohol, but dissolve readily in ether; they burn with a bright flame, and are not volatile. They are found as exudations on leaves and fruits, where they form the *glaucous* surface, which repels water. Some fruits, as the bayberry, are thickly coated with wax.

169. *Gums and Resins.* — In the strict sense, a *gum* is a substance which dissolves in water, forming a mucilage; but is insoluble in ether, alcohol, and oils. A *resin*,

on the other hand, is insoluble in water, and soluble in ether, alcohol, and oils. *Gum arabic* and *gum senegal* are products of different species of *acacia*, and are extensively used in the arts. *Gum tragacanth* is the product of a shrub which grows in Persia and Asia Minor. It is only partially soluble in water, forming a white paste instead of a mucilage. It is used for paste and cement, and also for stiffening woven fabrics. Dextrine is now much used as a substitute for these and other gums (162).

The *resins* are divided into the *hard resins*, the *soft resins*, and the *gum-resins*. The *hard resins* are brittle at ordinary temperatures, and are easily pulverized. The most important are *copal*, the varieties of *lac*, *mastic*, and *sandarach*. They are much used for varnishes and for other purposes in the arts. The *soft resins* are easily moulded by the hand, and some of them are semi-fluid, in which case they are called *balsams*. They consist essentially of hard resins dissolved in oils. On exposure to the air, they are oxidized and converted into hard resins. Under this head are placed *turpentine*, *Canada balsam*, etc. The *gum-resins* are the milky juices of certain plants, solidified by exposure to the air. They consist of a mixture of gums, resins, and essential oils. *Caoutchouc*, or *india rubber*, is the most important of the gum-resins. It is obtained from several quite different kinds of trees. The original *india rubber* of the East Indies is the product of a species of fig-tree. *Gutta-percha* is the dried milky juice of a tree found in Malacca and the neighboring islands. The applications of these two gum-resins in the arts are almost innumerable, and are continually increasing. Many valuable medicines are obtained from gums and resins. *Amber* is a *fossil* resin.

170. *Vegetable Alkaloids.* — A great variety of alkaline compounds, known as *alkaloids*, are produced by plants. They are mostly formed in the bark and in the

leaves. Among them are some of the most active poisons and some of the most valuable medicines, as *morphine*, *strychnine*, and *quinine*. Tea and coffee owe their peculiar effects to the same alkaloid, *theine*. *Opium* is remarkable for the great number of these compounds which it contains. The narcotic power of *tobacco* is due to a highly poisonous alkaloid called *nicotine*.

171. *Proteine Bodies.* — Under the term *proteine bodies* have been included the most important *nitrogenized* products of plants, as *albumen*, *caseine*, and *fibrine*. They all have essentially the same composition, containing about 53.6 per cent of carbon, 7.1 of hydrogen, 15.6 of nitrogen, 22.1 of oxygen, and a varying quantity of sulphur not exceeding 1.6 per cent. They are found in animals as well as in plants, and there are several similar compounds found only in animals.

The *proteine bodies* form the most essential articles of food for animals, since, as we have learned, *nitrogenized* compounds are indispensable to the repair of the machinery of the body. And for these, as for all other constituents of their food, animals are directly or indirectly dependent on plants.

SUMMARY.

The atmosphere contains oxygen, nitrogen, carbonic acid, and watery vapor, with traces of ammonia and nitric anhydride.

Ordinary combustion, decay, and respiration are chemical processes, differing from one another mainly in the rapidity and the completeness with which they take place.

The division of substances into *combustibles* and *supporters of combustion* is convenient, though merely conventional; since combustion consists in the chemical union of two or more elements, and, in reality, one of the elements is just as combustible as another.

The present materials of the earth, being mainly compounds of hydrogen and carbon with oxygen, and of the metals with oxygen, chlorine, and sulphur, must be regarded as products of combustion.

The *rusting* of the metals is a process of slow combustion.

The products of combustion, decay, and respiration are chiefly *carbonic acid*, *water*, and *ammonia*.

The growth of plants is a chemical process, carried on in the leaf by the agency of the sunbeam. In growing, plants remove from the air carbonic acid, water, and ammonia. These compounds are broken up; the oxygen in part given back to the air; and the other elements, with the remaining oxygen, re-arranged so as to form the various vegetable compounds.

By removing the products of combustion, and restoring oxygen, plants purify the air for the respiration of animals.

Animals derive all their food, directly or indirectly, from plants; while plants derive their food, directly or indirectly, from the air.

Most of the natural vegetable products are compounds of carbon, hydrogen, and oxygen; but some contain nitrogen in addition to these, and are, therefore, called *nitrogenized* compounds. Of the former class, the most important are starch, sugar, dextrine, the vegetable acids, the fats and oils, wax, the gums and resins, and the vegetable alkaloids; of the latter class, the proteine bodies, including albumen, caseine, and fibrine. The latter are the most essential articles of food for animals, since the nitrogenized products make up almost the entire bulk of the animal structure.

DESTRUCTIVE DISTILLATION AND ITS PRODUCTS.

172. When vegetable substances are heated in closed vessels, so as to exclude the oxygen of the air, they are broken up into a number of compounds, which vary with the temperature to which they have been exposed. When vegetable matter burns or decays, these organic compounds, as we have seen, are broken up by the affinity of oxygen brought to bear upon them from without; while in the other case they are broken up by the internal action of heat. In the first case, new compounds are formed by the addition of new material; in the second case, by subdivision, without any addition of new material. The first process is called *combustion;* the second, *destructive distillation.*

173. *The Preparation of Charcoal.*—We have a very simple case of destructive distillation in the preparation of charcoal. This process may be illustrated by putting pieces of dry wood into a flask, closed with a cork, through which a glass tube passes. This tube passes into a cold glass bottle, and thence to a jar over water. Heat the flask, and the wood turns black. The jar fills with an inflammable gas. Hold a cold glass vessel over the flame of this gas, and moisture collects upon it, showing that water is a product of the burning of this gas. Carbonic acid also is found to be produced. This gas obtained from the wood must,

therefore, be a compound of hydrogen and carbon. A part of the carbon of the wood combines with hydrogen, to form the inflammable gas; while the greater part remains behind as a black solid. This black residue is charcoal, a form of carbon. If the heating is continued, a liquid product is also obtained.

One way of preparing charcoal is to place billets of wood in an iron cylinder, which is closed air-tight and heated to dull redness. The volatile products are driven off and allowed to escape through a flue, and the solid charcoal remains behind.

A ruder method is practised in the country, where wood is plenty. A stake is set in level ground, and brushwood heaped about its base. Wood is then stacked round the stake, so as to form a mound some 20 or 30 feet in diameter. This mound is then covered, first with leaves or turf, and then with earth, leaving only a small opening at the bottom, through which the wood is set on fire. When the fire is well under way, the mound is covered more deeply and allowed to burn slowly out. This requires about a month. The burning of a part of the wood furnishes heat for charring the rest.

174. *The Products of the Distillation of Wood.*— When hard wood, as beech, is subjected to destructive distillation in a retort, and the volatile products are condensed in a suitable vessel, four principal classes of substances are formed: (1) gases; (2) a watery fluid; (3) a dark resinous fluid; (4) charcoal.

(1) This product is a mixture of inflammable gases, the most important of which are the two *hydrocarbons* (or compounds of hydrogen and carbon), *marsh-gas*, H_4C, and *olefiant gas*, H_2C.

(2) This product is an acrid liquid, known as *pyroligneous acid*, or *wood vinegar*, which is used in the manufacture of *acetic acid.*

(3) This product is *wood tar*, a thick liquid, insoluble in water, but soluble in alcohol. Its chief use formerly was for tarring and calking ships, but recently it has become an important source of illuminating and lubricating oils. These oils will be described hereafter.

(4) This product is the charcoal remaining in the retort. It is used chiefly as fuel and in reducing metallic ores.

175. *Ingredients of Wood Tar.*—When beech-wood tar is distilled, a light oil passes over first, called *eupion*, or *wood naphtha*. It is now often sold under the name of *benzole*, and used as a burning-fluid, for removing oil-stains from clothes, and for countless other purposes. It burns with a brilliant white flame, free from smoke; but its extreme inflammability makes it a dangerous liquid for lamps.

After this light oil has distilled over, a heavy oil follows. It contains various ingredients, the chief of which are *creosote* and *paraffine*.

176. *Creosote.*—This is an oily, colorless liquid, with a peculiar smoky odor. It has remarkable *antiseptic* (or preservative) properties. A piece of flesh steeped in a very dilute solution of it dries up into a mummy-like substance, which refuses to decay. Meat, as tongues or hams, may be almost instantly *cured* by dipping it into a solution containing one part of creosote to 100 parts of water or brine. It is this substance which imparts to wood-smoke its property of preserving meat. It is a compound of carbon, hydrogen, and oxygen.

177. *Paraffine.*—This is a pearly-white, tasteless, and odorless solid. The most corrosive acids and alkalies have no effect upon it. Hence its name, from *parum*, little, and *affinis*, from which *affinity* is derived.

It burns with a bright white flame, without smoke. It is now much employed as a material for candles, which for purity and lustre are not surpassed by even the best wax candles.

Unsized paper, after having been soaked in paraffine, may be kept for weeks in concentrated sulphuric acid without undergoing the slightest alteration. Hence it is an excellent coating for the labels of bottles in which acids are kept.

178. *Asphalt.*—Asphalt, or pitch, is the residue left after distilling tar. It is used for varnishes, and as a material for making lamp-black.

179. *Slow Destructive Distillation.*—When vegetable substances decay with a partial or complete exclusion of air, we have a kind of *slow* destructive distillation. Hydrocarbon gases, which in some places escape from the earth in large quantities, are one of the products of this process; the well-known coal-oils are another; and coal is a third. The gas obtained by stirring the mud in marshes and at the bottom of stagnant pools is formed in this way, and is made up chiefly of marsh-gas, H_4C, and carbonic acid, CO_2.

180. *The Formation of Mineral Coal.*—In tropical swamps where vegetation is rank, vast masses of vegetable matter accumulate, and gradually decay under water. In some cases the land at the bottom of these swamps is slowly sinking; and the bed of peat, as it sinks with it, becomes covered with mud and sand, which numerous streams are washing down upon it. This goes on, year after year and century after century, until the bed is buried hundreds of feet beneath the surface. The vegetable matter, thus sunk in the earth and subjected to enormous pressure, gradually undergoes a process of internal combustion similar to that which takes place under water. In many cases the decomposition is hastened by the agency of the internal heat of the earth. It is probable that the vast beds of coal found in various parts of

the earth have been thus formed. All this coal is the remains of an ancient vegetation, and it undoubtedly required millions upon millions of years to complete its conversion into coal.

181. *Hard and Soft Coals.* — The mineral coals may be conveniently divided into *hard*, or *anthracite*, and *soft*, or *bituminous* coal, and there are several varieties of each.

The main differences between the two are these: hard coal is almost pure carbon, while soft coal contains also considerable hydrogen and some oxygen; hard coal still retains the cellular structure of the wood, which is clearly seen under the microscope, while in soft coal this cellular structure is almost entirely wanting; hard coal burns without flame, soft coal with flame.

182. *Products of the Distillation of Bituminous Coal.* — The products obtained by the destructive distillation of coal are still more numerous than those obtained from wood. Wood, containing much oxygen and comparatively little nitrogen, furnishes compounds which contain much acetic acid and little ammonia, and which, therefore, have an *acid* reaction. Coal, on the other hand, contains much nitrogen and little oxygen, and gives products rich in ammonia, and having consequently an *alkaline* reaction.

When coal is distilled at high temperatures, an abundance of an inflammable gas is obtained, and also a large amount of liquid products, which are then called *tars*.

When coal is distilled at a low temperature, little gas is obtained, and the liquid products are then called *oils*.

183. *The Composition of Coal-tar.* — Coal-tar has been found to contain three classes of substances: —

(1) Acid oils.
(2) Alkaline oils.
(3) Neutral oils.

(1) The most important and abundant ingredient of the acid oils is *carbolic acid*, C_6H_6O. It is analogous to creosote, and has even more powerful antiseptic properties. It is one of the most valuable *disinfectants* known, and both the acid and the compound which it forms with lime are now much used for this purpose. It is also largely employed in dyeing silk and woollen goods. The dye-stuff is prepared by heating carbolic acid moderately with nitric acid, and is called *picric acid*. On evaporating this liquid, yellow scaly crystals are obtained. Like all the tar colors, its dyeing qualities, when in solution, are most intense. Silk and woollen goods put into the solution, even when cold, assume a rich yellow color, far surpassing that of other dyes.

(2) The alkaline oils constitute but a small fraction of the tar. Their most important ingredients are ammonia and *aniline*, C_6H_7N.

Aniline is an oily substance which, when acted upon by compounds which readily part with oxygen, furnishes a complete series of the most brilliant dyes. The preparation of these rich dyes from aniline is one of the most interesting discoveries of modern times, and has caused almost a revolution in the arts of dyeing and calico-printing. It is still more surprising when we consider that these brilliant colors are obtained from what was until recently a disagreeable waste product of the gas-works. When first prepared, they were worth their weight in gold; now, they can be bought at a comparatively moderate price.

Their dyeing qualities are so intense, that a little material goes a great way; so that, notwithstanding their high price, they are more economical than other dyes.

(3) The neutral oils are the *coal-oils* proper. They contain a great variety of compounds, both liquid and solid, the latter being held in solution. Of the liquids,

benzole, *toluole*, and *cumole* are the most important; and of the solids, *paraffine* and *naphthaline*.

184. *Benzole and Nitro-benzole.* — Benzole is a very important compound, as it is the material from which aniline is usually prepared. The symbol for benzole is C_6H_6. When mixed with concentrated nitric acid, it loses one atom of H, and takes one of NO_2 in its place, and becomes *nitro-benzole*, $C_6H_5NO_2$.

Nitro-benzole is the artificial *oil of bitter almonds*, and is much used in the art of perfumery. Its most important use, however, is in the preparation of aniline. When heated with acetic acid ($C_2H_4O_2$) and iron filings, it loses two atoms of oxygen and takes up two of hydrogen, and is converted into aniline, C_6H_7N.

185. *Toluole and Cumole.* — Toluole, C_7H_8, and cumole, C_9H_{12}, are the chief ingredients of the well-known illuminating or lamp oils obtained from coal.

186. *Naphthaline.*—Naphthaline is a beautiful, pearly-white solid. It is inflammable, but burns with a smoky flame and a disagreeable odor. Brilliant red and blue colors, rivalling those prepared from aniline, have lately been obtained from this solid.

When vegetable matter is distilled at a high temperature, benzole and naphthaline are formed in great abundance, with but small quantities of toluole, cumole, and paraffine. When, on the other hand, the distillation is conducted at a low temperature, toluole, cumole, and paraffine are formed in large quantities, with but little benzole and naphthaline.

COAL-OILS.

187. At the beginning of the present century, the means of lighting our dwellings consisted, in the main, of poor tallow candles and dim and dirty oil-lamps. On

the continent of Europe, whale-oil and other animal oils were costly, and resort was had to natural tar and bituminous slate, in order to obtain illuminating oils. For more than twenty years past, lamp-oils have there been extensively prepared from wood, rosin, and bituminous matter.

In Great Britain and in this country the manufacture of coal-oils is of much more recent growth, because the extensive whale-fisheries supplied all the wants of the market.

The manufacture of coal-oil was introduced into this country in 1853, and was at first confined to those districts where bituminous coal could be mined at a cheap rate.

Soon after this manufacture was established in this country, and after the value of coal-oils came to be fully recognized, attention was drawn to *petroleum*, or rock-oil, as a ready means of supplying these oils cheaply. On examination, this oil was found to be analogous, in its composition and its properties, to that obtained by the destructive distillation of soft coal and other bituminous substances.

188. *The Origin of Petroleum.*—All scientific men are agreed that the petroleum found in the earth results from the decomposition of organic matter, and nearly all are agreed that it results mainly from the decomposition of *vegetable* matter. It is, however, a disputed point, whether it results from the original decomposition of the vegetable substances, or from the action of the internal heat of the earth on the bituminous coal at a subsequent period.

It is probable that the petroleum now found in the earth is the product both of the original decomposition and of subsequent distillation. Petroleum is, however, rarely found in contact with bituminous strata of any kind. It is more often found in fissures in sand rocks; rocks in which no oil could ever have been generated, since whatever organic matter they might have contained was too much exposed to atmospheric oxygen to admit of its being *bitumenized*, or made bituminous. It is not only impossible that the oil could have originated in these sand rocks, or in the sandy shales which underlie them in the oil region in Western Pennsylvania; but it is most probable that the oil ascended from still lower rocks in the form of vapor, which condensed in the cavities above. Since, then, petroleum is seldom found where it originated, but, ordinarily, in cavities of rocks higher up, it seems probable that it is *mainly* the product of distillation.

The chemical conditions essential to the generation of oil have evidently existed over a very wide area; but the oil is found only where fissures exist in the rocks. These fissures serve two purposes: one, to give space for the formation and expansion of the hydrocarbon vapor; the other, to furnish receptacles for the condensed oils. These fissures must connect with the sources of the oil. If they have any outlets at the surface of the earth, by which the more volatile portions of the oil may escape as gas, the oil within them becomes thicker and heavier. Hence, as a general rule, the oil found near the surface is heavy, the cavities containing it being likely to have outlets. It may, of course, happen that a deeply-seated fissure has such an outlet.

189. *How Petroleum is obtained.* — The oil is obtained by piercing one of these cavities by a well. It often happens that the upper part of the cavity is filled with pent-up uncondensible gases. In this case, if the well happens to pierce the lower part of the cavity, the expansive force of the confined gases will drive the oil from the well in a continuous stream. Oil is often forced

from a new well with such velocity, that it rises in a jet a hundred feet high.

It sometimes happens that the lower part of the cavity is filled with brine, upon which the oil floats. If, in this case, the well pierces the lower part of the cavity, brine is the first product. After a time the salt-well may change to an oil-well.

COAL-GAS.

190. *The Manufacture of Coal-Gas.* — The idea of turning hydrocarbon gases to the practical purposes of illumination occurred at about the same time to Murdock, in England, and Lebon, in France. This was in 1790; but for a long time there was a strong popular feeling against the new light. In England, it was not until 1812 that Parliament consented to charter a company for its manufacture. Even then the enterprise was looked upon as so visionary, that the act of incorporation was said to be granted in order to make a great experiment of a plan of such extraordinary novelty. In December, 1813, Westminster Bridge was first lighted with gas. From this time gas-lighting made the most rapid progress in England; and now the consumption of gas in London alone amounts to more than seven billions of cubic feet annually. To make this gas, eight hundred thousand tons of coal are required; while the length of the main pipes in the streets of the city is more than two thousand miles.

Paris was first lighted with gas in 1820. There, as in England, strong prejudices had to be overcome.

The most essential parts of the apparatus used in the making of coal-gas are represented in Figure 28.

Of course, it is only soft or bituminous coal that can be used for making gas. This coal is distilled in long iron retorts, seen at the left of the figure. When charged with coal, these retorts are closed air-tight. They are

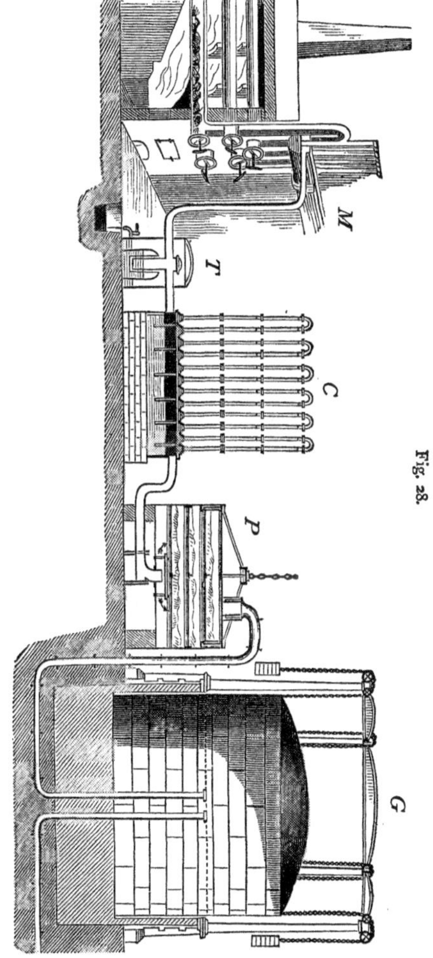

Fig. 28.

then heated to a very high temperature by the furnaces beneath.

The gaseous and volatile compounds, formed by the distillation of the coal, pass up through pipes (one of which may be seen leading from each retort) into a long horizontal pipe, *M*, called the *hydraulic main*. This is half full of water, into which the pipes from the retorts dip. The gas readily escapes from the pipes by bubbling up through the water; but, of course, it cannot return through the water.

The gas is carried from the hydraulic main through a pipe into a tank, *T*, called the *tar cistern*. By this time the more condensible gases have become liquid, and collect in the smaller vessel into which the pipe passes, and from which they overflow into the larger tank. From the latter they are drawn off at intervals. These condensed products are coal-tar and a liquid highly charged with ammonia, called *ammoniacal liquor*.

The uncondensed gases pass on through the series of upright pipes at *C*, called the *condenser*. Here they become still further cooled, and all the remaining condensible gases are reduced to the liquid state.

After leaving the condenser, the gas still contains, besides the compounds fit for illuminating purposes, the noxious compounds carbonic acid and hydric sulphide (49). These are removed in the *purifier*, *P*. It consists of a chamber with several perforated shelves, which are covered with slaked lime. In passing over this lime, the carbonic acid and hydric sulphide are absorbed, while the purified gas passes along into the *gasometer*, *G*.

This gasometer consists of a large sheet-iron bell-jar, which dips into a cistern of water. The latter is deep enough to allow the bell to be completely submerged, and filled with water. The bell is counterpoised with weights, and rises as the gas enters it.

From the gasometer the gas passes out, by the pipe at the right, into the streets and houses of the city.

ILLUMINATION.

191. *Nature of Flame.* — We know that coal-gas burns with flame, while the solid carbon burns without flame. All combustible gases burn with flame.

We have seen that wood, when heated in a closed tube, gives off an inflammable gas; and the same is true of wax or tallow heated in the same way. Every *solid* which appears to burn with flame can thus be converted into an inflammable gas by means of heat. When these solids begin to burn, the heat developed is sufficient to generate this inflammable gas. It is this gas, and not the solid, which burns with flame.

Flame, then, is gas burning.

192. *A Solid and Heat are necessary to Illumination.* — The oxy-hydrogen flame (22) develops intense heat, but scarcely any light. If, however, a cylinder of lime be held in this flame, we get the brilliant *calcium*, or *Drummond* light.

We see, then, that a solid and an intense heat are necessary to illumination.

193. *These two Conditions are fulfilled in the Burning of Coal-Gas.* — Illuminating gas, as we have seen, is a compound of carbon and hydrogen. If we hold a cold glass rod in a gas-flame, it becomes blackened with soot, showing that there is free carbon in the flame. We therefore conclude that the hydrogen of the gas combines with the atmospheric oxygen first. In doing so, it develops great heat.

In the manufacture of charcoal, the delicate cells of the wood are preserved intact; showing that the carbon, at the high temperature to which it is exposed in the process, has no disposition to melt. Its infusibility is its most marked characteristic (41).

When the hydrogen of coal-gas combines with the oxygen of the air, the carbon is set free in a solid state. As the hydrogen consumes all the oxygen in the nearest layer of air, the carbon must delay a moment in the flame before it can get at oxygen to unite with. This finely divided solid thus becomes heated white-hot, and gives out light. This condition lasts but an instant. The next instant the carbon combines with atmospheric oxygen, and passes off as a colorless gas. As fast, however, as the carbon is consumed, a fresh supply is set free, and thus the light is constantly kept up.

The value, then, of coal-gas for illumination depends upon a most delicate adjustment of affinities.

Were the affinity of carbon for oxygen a little stronger than it now is, the carbon would burn at the same time with the hydrogen, and there would be no light in the flame. On the other hand, were the affinity of carbon for oxygen less than it is, the carbon would not burn at all, but, after developing light, would pass away from the flame in the form of soot.

194. *Bunsen's Lamp.* — The fact that carbon must remain an instant in the flame in the solid state to develop light, is illustrated by Bunsen's burner. It is shown in Figure 29, and consists of a brass tube, near the bottom of which are four round holes, through which the air can pass into the tube. A second tube opens into the inside of the brass tube from the bottom, just on a level with these holes. The coal-gas passes into the larger tube, through this second tube. The air passes in through the holes at the same time, and the two gases become intimately mixed before they reach the top of the tube. Upon lighting the mixture as it escapes from the tube, it burns

Fig. 29.

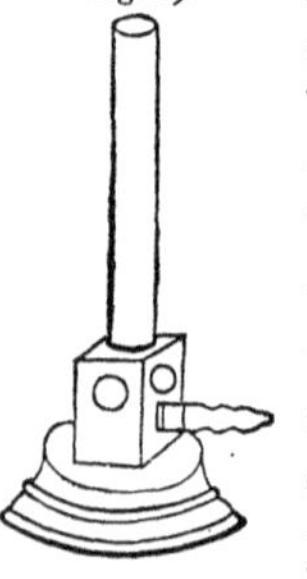

with scarcely any light, but a good deal of heat. If we close the holes at the bottom of the tube, the gas burns with the ordinary luminous flame. In the first case, there is an excess of oxygen mixed with the coal-gas, so that there is enough to combine with the carbon as soon as it is set free, before it becomes sufficiently heated to develop light. In the second case, the carbon cannot get at the oxygen to combine with it, until it has passed through the burning layer of hydrogen, and thus become intensely heated.

195. *The Best Shape for a Gas-Flame.* — From what has been said, it is evident that the light of the flame is developed only at the surface, in the burning layer of hydrogen. The ordinary gas-burner is so made that the flame will be flat, in order that it may have as much surface as possible. The greater the surface, the greater the light developed.

196. *The Argand Burner.* — A sheet of paper will evidently present to the air just as much surface when it is bent round till the two edges meet, as when it is flat; while in the second case it is in a more compact form. So the ordinary gas-flame will present just as much surface, if its edges are bent round till they meet. In the Argand burner, this form is given to the flame, by supplying the gas through a circle of small holes in a hollow brass ring. A current of air passes up through this ring, furnishing oxygen to the inner surface of the flame.

197. *The Bude Light.* — If the interior of the ring of an Argand burner be closed, the flame becomes of a dull red color. Since no oxygen is supplied to the inner surface of the flame, the combustion is imperfect, and develops but a low degree of heat.

If, on the other hand, a stream of pure oxygen be supplied to the interior of the cylindrical flame, the flame diminishes in size; but the light becomes very intense, and of a pure white color. The increased supply of oxygen increases the energy of the combustion of the hydrogen, and the intensity of the heat developed by it. In the first case, the carbon is heated only to a dull redness, while in the second case it is raised to a full white heat.

198. *A Burning Candle is a miniature Gas-Factory.* — We have seen that wax or tallow, when heated in a flask, is converted into an inflammable gas. This gas, in burning, produces water and carbonic acid; hence it must contain hydrogen and carbon, which are the main ingredients of coal-gas. This gas is in fact identical with coal-gas. The heat of the burning match converts some of the wax of the candle into gas, which takes fire, and in burning develops sufficient heat to convert more of the wax into gas; and thus the supply is kept up.

199. *The Flame of a Candle consists of three distinct Parts.* — In the flame of a candle there are three distinct parts: (1) the dark central zone, or supply of unburnt gas surrounding the wick; (2) the luminous zone, or area of incomplete combustion; and (3) the non-luminous zone, or area of complete combustion. If we put one end of a small glass tube (see Figure 30) into the dark central zone, the unburnt gas will pass up the tube, and may be ignited at the other end. In the luminous part of the flame, as in the case of the gas-flame (193), the gas is not completely burnt, and carbon is separated in the solid state; and it is this carbon heated white-hot which renders the flame luminous. In the outer zone the supply of oxygen is greater; all the carbon is at once burnt to carbonic acid, and the flame here becomes non-luminous.

Fig. 30.

SUMMARY.

Vegetable compounds may be broken up by *destructive distillation.*

In the preparation of charcoal, wood is decomposed into carbon and a number of volatile compounds.

When these volatile products are cooled and condensed, we obtain: (1) permanent gases; (2) a watery fluid; (3) a dark resinous fluid.

(1) The permanent gases are chiefly *marsh-gas*, H_4C, and *olefiant gas*, H_2C.

(2) The chief ingredient of the watery fluid is *pyroligneous acid*, or *wood vinegar.*

(3) The dark liquid is *wood tar.*

The most important ingredients of wood tar are *wood naphtha*, *creosote*, *paraffine*, and *asphalt.*

Mineral coal has probably been formed by the gradual decomposition of vegetable matter buried in the earth, and thus excluded from the air; the decomposition in many cases being hastened by the internal heat of the earth.

When the volatile products were allowed to escape freely, *anthracite* coal was formed; when they could not escape, *bituminous* coal was formed.

The volatile products often escaped into large cavities in the rocks, and on condensing gave rise to the celebrated *rock-oil*, or *petroleum.*

In some cases, this oil may have been formed by the decomposition of the vegetable matter at a temperature too low for its conversion into vapor, and may afterwards have been distilled by the internal heat of the earth.

The products of the destructive distillation of bituminous coal are more numerous than those of the distillation of wood. They are less rich in compounds containing *oxygen*, but richer in those containing *nitrogen.*

When the coal is distilled at a high temperature, a large amount of gas is obtained, and the liquid products are then called *tars*. When the coal is distilled at a low temperature, only a small amount of gas is obtained, and the liquid products are called *oils*.

When the gas obtained by the distillation of coal is purified, it forms the well-known *illuminating gas*.

The most important ingredients of *coal-tar* are *carbolic acid*, *ammonia*, *aniline*, *benzole*, *toluole*, *cumole*, *paraffine*, and *naphthaline*.

The *aniline colors* are prepared chiefly from benzole.

Only gases burn with flame. Solids that appear to burn with flame are first converted into a gas by heat, and this gas burns with flame.

The two conditions essential to illumination are the presence of a *solid* and intense *heat*.

The illuminating gases are all *hydrocarbons*, or compounds of hydrogen and carbon.

The hydrogen has the stronger affinity for oxygen, and burns first. The carbon is set free in a solid state, and delaying for an instant in the intense heat of the flame, before combining with oxygen, becomes luminous.

The light of the flame is increased (1) by increasing the surface of the flame, as in the ordinary gas-flame, and that of the Argand burner; and (2) by increasing the intensity of the combustion of the hydrogen, as in the Bude light.

The heat of the flame is increased by mixing oxygen with the gas before it burns, so that the carbon and hydrogen may burn together, as in the Bunsen burner.

The burning candle or oil-lamp is a miniature gas-factory.

The candle flame consists of three parts: (1) an inner zone of unburnt gases, surrounded by (2) a luminous zone of incomplete combustion, outside of which is (3) a non-luminous zone of complete combustion.

FERMENTATION.

200. *Causes of Fermentation.* — Complex organic compounds are often broken up into simpler compounds by the agency of growing plants. This process is called *fermentation*, and the organisms which effect it are called *ferments*. Different ferments give rise to different products, as alcohol, vinegar, etc. Most of these ferments are plants, but one at least is an animal; and this, strange to say, cannot live in contact with free oxygen, but flourishes in an atmosphere of hydrogen. In order that the ferment should grow, it must be supplied with proper food, especially with ammoniacal salts and alkaline phosphates. These are contained in the albuminous matter generally present in the liquid about to be fermented. In order that the fermentation should go on well, the temperature should be from 70° to 100°; at much higher, as at much lower temperatures, the vitality of the ferment is destroyed.

In many cases, spontaneous fermentation sets in without the apparent addition of any ferment: thus wine, beer, milk, etc., when allowed simply to stand exposed to the air, become sour or otherwise decompose. These changes are, however, not effected without the presence of vegetable or animal life, and are true fermentations. The *sporules*, or seeds of these living bodies, are always floating in the air, and on dropping into the liquid begin to propagate themselves, and in the act of growing evolve the products of the fermentation. If the liquid be left only in contact with air which has been passed through a red-hot platinum tube, so that the living sporules are destroyed; or if the air be simply filtered by passing through cotton-wool, so that the sporules are prevented from coming into the liquid, it is found that the liquid may be preserved for any length of time without undergoing the slightest change.

The principal forms of fermentation are—

(1) The *alcoholic* fermentation, producing chiefly alcohol and carbonic acid.

(2) The *acetous* fermentation, producing acetic acid.

201. *Alcoholic Fermentation.* — Sugar, if dissolved in presence of yeast (which is a plant and a ferment), undergoes fermentation, evolving mainly alcohol and carbonic acid:—

$$C_{12}H_{24}O_{12} = 4C_2H_6O \text{ (alcohol)} + 4CO_2.$$

About 6 per cent of the sugar undergoes a different change, part being used as nourishment for the yeast, and part forming glycerine and other compounds. The alcoholic fermentation is best effected at a temperature between 77° and 86°.

When the cereal grains are used for making alcohol, the starch of the grain is first converted into sugar. This change may be brought about by the action of sulphuric acid, as already explained (161); but, practically, it is usually effected by means of *diastase.*

202. *The Formation of Diastase in Growing Plants.* — We have learned that the seed contains a supply of starch which is the food of the embryo plant until it is able to derive its sustenance from the earth and the air. We have learned, too, that this starch must be converted into sugar, in order that it may be dissolved in the sap of the plant and carried where it is needed for the purposes of growth. Let us now see how the starch is changed to

sugar. The seed contains more or less of the *nitrogenized* compound, *gluten* (142). This, under the influence of heat and moisture, begins to putrefy, and a portion of it is converted into the ferment *diastase*. This has so powerful an action upon starch, that one part of it, at a temperature of 150°, is sufficient to convert 2000 parts of starch into dextrine and then into sugar.

Fig. 31.

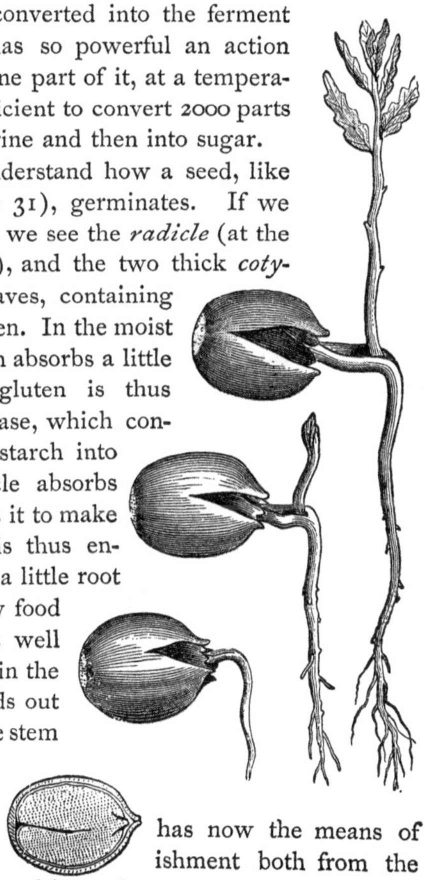

We can now understand how a seed, like the acorn (Figure 31), germinates. If we cut the acorn open, we see the *radicle* (at the right in the figure), and the two thick *cotyledons*, or seed-leaves, containing the starch and gluten. In the moist warm soil the acorn absorbs a little water, and the gluten is thus changed into diastase, which converts some of the starch into sugar. The radicle absorbs this sugar, and uses it to make woody fibre. It is thus enabled to thrust out a little root and begin to draw food from the earth, as well as from the starch in the seed. It next sends out the *plumule*, a little stem with its first leaves unfolding to the air and light. It has now the means of deriving its nourishment both from the earth and the air, and is no longer dependent upon the seed, which soon withers and perishes. Both root and stem are now rapidly developed (145–154), and thus at length the acorn becomes the mighty oak.

203. *Brewing.*—The brewer avails himself of this natural change in the germinating seed, and calls into action on a larger scale the chemical influence of diastase. The grain commonly used for the purpose is barley. This is first moistened in heaps, and spread out in a dark room to heat and sprout. When the gluten has begun to be transformed into diastase and the starch into sugar, so that the germ is about to burst from the envelop of the seed, the growth is arrested by heating the grain, and thus killing the embryo plant.

The *malt*, as the barley is now called, is next bruised, and soaked in warm water in the *mash-tub*. The water dissolves first the sugar which has already been formed in the seed, and then the diastase. The latter acts upon the rest of the starch of the seed, and of any raw grain which may be added to the malt, converting it into sugar, so that little, except the husk of the grain, is left undissolved; and the liquor, or *wort*, has a decidedly sweet taste.

The wort is now heated to boiling, to stop the action of the diastase, and to coagulate the albumen which has been dissolved out of the grain. At the same time, hops are added to the wort; and these, besides giving a bitter aroma to the liquid, help to clarify it. The boiled liquor is then filtered, and drawn off into shallow vessels, where it is cooled down to about 60°. Yeast is now added, and the mixture is allowed to ferment for six or eight days.

This fermentation is never allowed to continue until all the sugar is converted into alcohol. From one-half to three-fourths of it is decomposed; and the rest is left to give a sweet taste to the beer, and also to keep it from souring in the cask.

The liquor is next put into casks, where it goes through a second and much slower fermentation, which is essential to its preservation. When this fermentation is completed, the cask must be closed tight, to exclude the air, the oxygen of which would cause *acetous* fermentation. During this second fermentation, the carbonic acid is generated, to which the liquor owes its sparkling *effervescent* character.

204. *The Distillation of Spirits.* — When fermented liquors are boiled, the alcohol they contain rises in the form of vapor, mingled with steam. If the boiling be performed in a close vessel, and the vapors be led by a pipe into a cold receiver, they condense into a liquid, containing a large percentage of alcohol. This process is called *distillation*, and the apparatus for carrying it on is called a *still.*

If the alcoholic liquor thus obtained be re-distilled, or *rectified*, a spirit may be obtained, containing 90 per cent of alcohol and 10 of water. This is the strongest commercial alcohol. If this be mixed with calcic chloride, which has a strong affinity for water (113), and then be re-distilled, pure or *absolute* alcohol may be obtained.

We have seen that, in making beer, the fermentation of the wort is checked before all the sugar is converted into alcohol. But, in the manufacture of spirits from grain, the fermentation is prolonged until all the sugar is transformed into alcohol and carbonic acid. To leave any of it unchanged, would not only involve a loss of spirit, but, during the subsequent distillation, might injure the flavor of the spirit obtained.

The alcohol obtained from the fermentation of sugar is sometimes called *wine alcohol*, or *vinic alcohol*, to distinguish it from a large number of similar compounds, known to chemists as *alcohols.* Its symbol is C_2H_6O.

205. *Ether.* — When equal weights of strong alcohol and oil of vitriol are boiled in a retort, a colorless, fragrant, and highly volatile liquid passes over, called *ether*, or *sulphuric ether*, $C_4H_{10}O$. It is called *sulphuric*

ether, to indicate its origin, and to distinguish it from other analogous compounds called *ethers*.

When poured on the hand, ether produces a strong sensation of cold, owing to the rapidity with which it evaporates, rendering heat latent. It is very inflammable, and burns with a white flame. Its vapor forms explosive mixtures with air or oxygen. It dissolves oils, fats, and resins far more readily than alcohol does. It also dissolves sulphur, phosphorus, and iodine. When inhaled in sufficient quantity, its vapor is an *anæsthetic*, and was the first substance used for that purpose by physicians.

Acetic ether is obtained by distilling a mixture of alcohol, sulphuric acid, and acetic acid. It is used in medicine. If nitric acid be used, we get *nitric ether*, which is well known under the name of *sweet spirits of nitre*. Several of the ethers exist in a natural state in fruits, giving them their peculiar flavors.

206. *Acetic Acid.* — If alcohol, diluted with water, be mixed with a ferment, as yeast, and exposed to the air at ordinary temperatures, it is soon converted into *acetic acid*, or *vinegar*. The alcohol absorbs 2 atoms of oxygen, forming water and acetic acid: —

$$C_2H_6O + O_2 = H_2O + C_2H_4O_2 \text{ (acetic acid).}$$

This process constitutes what is commonly called the *acetous fermentation*. Since the alcohol is oxidized, it is in a certain sense a *combustion* that takes place. It is, however, brought about by the influence of a ferment, known as the *vinegar-plant*, or *mother of vinegar*.

There are a great many methods of making vinegar on the large scale; but, for the most part, they are only different devices for exposing to the air as large a surface as possible of the alcoholic mixture. In one process, the liquid is allowed to trickle down through vats, filled with flat pieces of wicker-work, piled one upon another. These are first watered with vinegar, or liquid partially converted into vinegar, and thus become covered with the germs of the vinegar-plant. Air is admitted through holes in the bottom of the vats, and passes out at the top, after giving up a part of its oxygen to the alcohol. Heat is produced by the oxidation, and this causes an upward draught, and quickens the circulation of air through the vats.

In another process, the weak alcoholic liquor trickles down through perforated casks, filled with beech shavings, which have been thoroughly scalded to remove all soluble matter. Air is admitted by holes around the lower part of the cask, and passes out at the top.

When vinegar is purified, and concentrated by distillation, it constitutes the *acetic acid* of commerce. This is prepared on a large scale from *pyroligneous acid*, or *wood vinegar* (174), which is obtained by the destructive distillation of wood.

207. *Bread-making.* — The making of bread in the ordinary way, by means of leaven or yeast, is an example of alcoholic fermentation.

When the grain of wheat is ground in a mill, and then sifted, it is separated into two parts, the *bran* and the *flour*. The bran is the outside harder part of the grain, which is not ground so readily, and, when ground, makes the flour darker. Both the bran and the flour consist mainly of gluten, starch, and water.

If the flour be mixed with water enough to moisten it thoroughly, the particles cohere and form a *dough*, which can be kneaded or moulded with the hand. It is the gluten of the flour which gives the tenacity to the dough.

A little leaven or yeast is added to the flour, either before it is mixed with water into a dough or in the course of this process; and the dough is then placed for some hours in a warm atmosphere, in order that it may *rise*. This rising is a fermentation, caused by the leaven or yeast. Leaven is the primitive ferment, and is simply a piece of moistened dough which has begun to putrefy. When brought in contact with a fresh portion of flour and water, it very quickly acts as a ferment, and develops fermentation in the whole mass. Yeast, as has been stated, is one of the vegetable growths which are ferments (200, 201).

The leaven or yeast soon begins to act on the gluten, starch, and sugar of the flour, and a portion of the sugar is converted into carbonic acid and alcohol in every part of the dough. The bubbles of gas, thus disengaged in the mass, form innumerable cavities, and make it light and porous.

The spongy dough is now put into a hot oven, where the fermentation and swelling are at first increased by the heat; but when the whole has been heated to about 212°, the fermentation is suddenly arrested, and the mass becomes fixed in the form it has then attained. In the baking, some of the water is dissipated from the dough, the starch and gluten are partially *boiled*, and some of the starch is converted into dextrine. The brown, glossy appearance of the crust is due to this formation of dextrine (162). Most of the alcohol is driven off by the heat; but a large amount of water (about 45 per cent of the weight of the bread) remains in the loaf after the baking.

There are various methods of making bread without fermentation, most of which depend on the liberation of carbonic acid from one of its compounds, by means of an acid. Bicarbonate of soda and muriatic acid are the materials sometimes used. Cream of tartar, which is a compound of tartaric acid (165), is often used as the acid.

The salt and acid are usually mixed with the dough, and the carbonic acid there set free; but in the manufacture of the so-called *aerated bread*, water charged with the gas is used in making the dough.

SUMMARY.

Fermentation is the breaking up of complex organic compounds into simpler ones, by the agency of certain plants called *ferments*.

The *alcoholic* fermentation takes place in sugar, and produces alcohol and carbonic acid.

In making alcoholic liquors from grain, the starch of the grain is converted into sugar, by the action of *diastase*, as in the germination of seeds. The sugar is then converted into alcohol by the use of a ferment, as yeast.

The alcohol of fermented liquors may be separated from the water, and concentrated by *distillation*.

By the action of acids upon alcohol, important compounds called *ethers* are obtained.

The *acetous* fermentation takes place in alcohol, which undergoes oxidation, producing vinegar, or acetic acid.

The making of bread with yeast is an example of alcoholic fermentation.

APPENDIX.

CHEMICAL PHILOSOPHY.

ATOMS AND MOLECULES.

1. *Molecular and Atomic Constitution.* — The astronomer finds that the universe is made up of stars, or suns, each of which is probably, like our own sun, the centre of a solar system made up of planets. These solar systems are the larger units, or individuals, of the universe; and the planets are the smaller units of which these systems are composed. Some of these smaller units, as Venus and Mars, are simple; others, as the Earth and Jupiter, are complex, being made up of a planet and one or more moons. We now believe that all the bodies about us are made up of units analogous to these. The physicist finds matter to be made up of separate and distinct masses, or units, which he calls *molecules*. He cannot see these even with the aid of the microscope; but he is as certain of their existence as the astronomer is of his planets and solar systems. The chemist pulls these molecules in pieces, and finds that they are made up of smaller units, which he calls *atoms* or *compound radicals*, according as they are simple or complex. He believes that in most cases the molecules of the elements, as well as of compounds, are made up of two or more atoms; the only difference being, that in the one case the atoms are all alike, while in the other they are not. Thus a molecule of hydrogen is believed to be made up of two atoms of hydrogen; while a molecule of water is made up of two atoms of hydrogen and one of oxygen. When a molecule of water is broken up, it ceases to be water, and becomes hydrogen and oxygen. Hence we may

define a *molecule* as *the smallest mass of a substance which can exist by itself;* and an *atom* as *the smallest mass of any substance which can exist in a molecule.*

The molecules of a body are bound together by *cohesion*, and the atoms by *affinity*. The *state* of a body depends upon the strength of the cohesive force. When this force binds the molecules together firmly, they form a *solid;* when very weakly, a *liquid;* and when not at all, a *gas*. The molecules of a solid, as a general rule, tend to arrange themselves in regular order, so as to build up symmetrical forms called *crystals*. These crystals have usually the same form for the same substance, but very different forms for different substances.

It is probable that the *atoms* when bound together by *affinity* tend to arrange themselves with equal regularity; but as yet we are ignorant of the mode of arrangement. Indeed, the *number* of atoms in the molecules of many substances is as yet a matter of mere hypothesis.

2. *Molecular Weight.* — When the molecules of any body are so far apart that the cohesive force exerts no constraint upon them, the substance is called a *true gas*. When the molecules are still somewhat under the restraint of cohesion, but not bound together by it, the substance is called an *imperfect gas*, or *vapor*. Now there are two characteristics of all true gases, which throw light upon their molecular constitution. "First, all true gases obey the same law of compressibility. Secondly, equal volumes of all true gases expand equally on the same increase of temperature." The repulsive force, then, which acts among the molecules, and tends to keep them apart, or to separate them, is the same for all true gases. This has led to the conclusion that *equal volumes of all true gases have the same number of molecules*. We can therefore readily find the *molecular weight* (that is, *the relative weight of their molecules*) of all those substances which can by heat be converted into true gases, by weighing equal volumes of these gases under the same pressure and temperature.

There are, however, many substances which cannot be brought into a gaseous state. To determine the molecular weights of these substances, we must find their atomic constitution, and the atomic weights of their elements. Suppose, for instance, that we do not know the molecular weight of *water*, but do know that a molecule of water contains two atoms of hydrogen and one of oxygen, and that the atom of hydrogen weighs 1 and the atom of oxygen 16. The molecular weight of water must then be 16 + 2 or 18.

In finding the molecular weight of different substances, they are all compared with *hydrogen*, whose molecular weight is assumed to be 2. The following table gives the molecular weights of several gases, all of which contain hydrogen; also the relative weight of the hydrogen in each molecule: —

	Molecular Weight.	Weight of Hydrogen.		Molecular Weight.	Weight of Hydrogen.
H	2	-	H_2S	34	2
HCl	36.5	1	H_3N	17	3
HBr	81	1	H_3P	34	3
HI	128	1	H_3As	78	3
H_2O	18	2	H_4C_2	28	4
			H_4C	16	4

3. *Atomic Weights.* — When any element forms a number of compounds which can be brought into the gaseous state, we find its atomic weight by *first finding the molecular weight of these compounds*, and *then analyzing these compounds, so as to find the weight of the common element in each molecule*, or, what amounts to the same thing, *in equal volumes of the gases.*

Thus, in the above table, it is found that when HCl is decomposed, it gives *half its own volume*, or, in other words, *half its own number of molecules*, of hydrogen. Hence a molecule of HCl contains *half a molecule*, or *one part by weight*, of hydrogen. The same is found to be true of HBr and HI. *Water* gas is found on analysis to give *its own volume*, or, in other words, *its own number of molecules*, of hydrogen. Hence each molecule of water contains *a whole molecule*, or *two parts by weight*, of hydrogen. The same is found to be true of H_2S. Again, H_3N gives *one and a half its own volume* of hydrogen. A molecule of ammonia, then, contains *one and a half molecules*, or *three parts by weight*, of hydrogen. The same is true of H_3P and H_3As. Again, H_4C (marsh-gas) gives *twice its own volume* of H; and each of its molecules must therefore contain *two molecules*, or *four parts by weight*, of H.

In this way, it is found that *one part by weight* is the smallest quantity of hydrogen found in a molecule of any of its compounds. Hence 1 is taken as the atomic weight of hydrogen; and since the molecular weight of hydrogen has been assumed as 2, each of its molecules must contain 2 atoms.

In a similar way, it has been found that the molecular weight of nitrogen is 28, and its atomic weight 14. "This method of investigation can be extended to a large number of the chemical elements, and the conclusions to which it leads are evidently legitimate, and cannot be set aside until it can be shown that some substance exists whose molecule contains a smaller mass of any element than that hitherto assumed as the atomic weight, or, in other words, until the old atom has been divided."

This method is, however, applicable only to those elements which form a number of gaseous or volatile compounds. In other cases, the number of atoms in a molecule is a matter of inference merely. Whenever analogous elements, as iodine, bromine, fluorine, and chlorine, combine with the same element, we infer that the molecules formed have the same atomic constitution, especially if the compounds have the same crystalline form and similar chemical relations. Of course, if we know the atomic *constitution* of a compound, and the atomic *weight* of one of its elements, we can find the atomic weight of its other elements.

Suppose, for instance, we know that the molecule of *baric chloride* contains one atom of barium and two of chlorine, and that the atomic weight of chlorine is 35.5, we find on analysis that baric chloride contains 137 parts by weight of barium to 71 of chlorine, $71 = 35.5 \times 2 =$ the weight of two atoms of chlorine. Hence the weight of the one atom of barium must be 137.

"Nevertheless, it is true in very many cases that our conclusion in regard to the number of atoms which a molecule may contain is more or less hypothetical, and hence liable to error and subject to change. This uncertainty, moreover, must extend to the atomic weights of the elements, so far as they rest on such hypothetical conclusions."

Experiment, however, has shown that it is *generally* true that the elementary atoms have the same *specific heat;* that is, if we take of different elements quantities proportionate to their atomic weights, and which therefore contain the same number of atoms, ***the same amount of heat will raise them all the same number of degrees in temperature;*** in other words, it takes just as much heat to raise 1 atom of hydrogen 1 degree in temperature as it does to raise 1 atom of nitrogen, oxygen, or any other element 1 degree.

Hence, when we are in doubt as to the atomic weight of any element, we can sometimes settle the question by its *specific heat.*

QUANTIVALENCE AND ATOMICITY.

4. *Quantivalence.* — When the atoms of different elements replace one another in a compound molecule, it is found that while in some cases 1 atom of an element may replace 1 of another, in other cases it may replace 2, 3, or 4 atoms of another. Thus, when hydric chloride and argentic nitrate act upon each other, we have the following change: —

$$AgNO_3 + HCl = AgCl + HNO_3.$$

Here 1 atom of Ag changes place with 1 of H.

When Zn acts upon H_2SO_4 we have

$$H_2SO_4 + Zn = ZnSO_4 + H_2.$$

Here 1 atom of Zn replaces, and is equivalent to, 2 atoms of H.

Again, in the reaction between water and phosphorous chloride, we have —

$$3H_2O \text{ or } H_3H_3O_3 + PCl_3 = H_3PO_3 + 3HCl.$$

Here 1 atom of P replaces, and is equivalent to, 3 atoms of H.

"This relation of the elements to each other is called by Hofmann *quantivalence;* and selecting here, as in the system of atomic weights, the hydrogen atom as our standard of inference, the atoms of different elements are called *uni*valent, *bi*valent, *tri*valent, or *quadri*valent, according as they are, in the sense already indicated, *equi*valent to 1, 2, 3, or 4 atoms of hydrogen.

These terms are very appropriate, since they are all derived from the same root as our common English word *equi*valent, which best expresses the fundamental idea which underlies the whole subject."

The *quantivalence*, whenever important, may be indicated by a Roman numeral placed over the atomic symbols; thus, —

$$\overset{\text{I}}{\text{Cl}},\quad \overset{\text{II}}{\text{O}},\quad \overset{\text{III}}{\text{N}},\quad \overset{\text{IV}}{\text{C}}.$$

5. *Atomicity.* — "The quantivalence of an element is shown, not only by its power of replacing hydrogen atoms, but also by its power of replacing any other atoms whose quantivalence is known. Moreover, what is still more important, the quantivalence of an element is shown not only by its replacing power, but also by what we may term its *atom-fixing power*; that is, by its power of holding together other elements in a molecule. We may take as examples the molecules of four very characteristic compounds; namely, hydrochloric acid, water, ammonia, and marsh-gas," whose symbols are $H\overset{\text{I}}{Cl}$, $H_2\overset{\text{II}}{O}$, $H_3\overset{\text{III}}{N}$, and $H_4\overset{\text{IV}}{C}$. "By these symbols it appears that, while the univalent atom of chlorine can hold but 1 atom of hydrogen, the bivalent atom of oxygen holds 2, the trivalent atom of hydrogen 3, and the quadrivalent atom of carbon 4 atoms of the same element."

"The quantivalence of the chemical elements, especially as indicated by their atom-fixing power, is by no means always the same. That of the univalent elements alone is constant. They always combine with single atoms. . . . But the other elements frequently exhibit, under different conditions, an unequal atom-fixing power. Thus we have $\overset{\text{II}}{Sn}Cl_2$ and $\overset{\text{IV}}{Sn}Cl_4$, $\overset{\text{III}}{P}Cl_3$, and $\overset{\text{V}}{P}Cl_5$, $\overset{\text{III}}{N}H_3$ and $\overset{\text{V}}{N}H_4Cl$. Each element, however, has a *maximum* power, which it never exceeds. This we shall call its *atomicity*; and we shall distinguish the elements as *monads*, *dyads*, *triads*, etc., according to the number of univalent atoms they are able at most to bind together. Thus nitrogen is a *pentad*, although it is commonly *tri*valent; and lead is a *tetrad*, although it is usually *bi*valent."

The prevailing quantivalence of an element is a more characteristic property than its atomicity. The former is generally well established, while the latter is frequently still in doubt.

6. *Quantivalence of Radicals.* — "When in the molecule of any compound the dominant or central atom is united to as many other atoms as it can hold of that kind, the molecule is said to be *saturated*. Thus HCl, H_2O, H_3N, and H_4C are all saturated molecules. On the other hand, the molecules $\overset{II}{C}O$, $\overset{II}{P}Cl_3$, and $\overset{II}{Sn}Cl_2$ are not saturated; for they can combine directly with more oxygen or chlorine, forming thus the saturated molecules CO_2, PCl_5, and $SnCl_4$." It will be seen that CO, PCl_3, and $SnCl_2$ are *bi*valent radicals, since they can each hold the equivalent of two atoms of hydrogen. The CO can hold an additional atom of oxygen, and PCl_3 and $SnCl_2$ can each hold two additional atoms of chlorine. The quantivalence of an atom of C is 4, and that of an atom of O is 2; and the difference of these is 2, the quantivalence of the radical CO. In general, the quantivalence of a binary radical is *the difference of the quantivalences of its elementary atoms*. Thus, the quantivalence of N is 5, and that of O_2 is 4; and that of the radical NO_2 is 1. The quantivalence of SO_2 is 2, that of S being 2, and that of O_2 being 4. The quantivalence of PO is 3, that of P being 5, and that of O being 2.

The quantivalence of ClO_2 is 1. This radical illustrates the way in which elements often seem to combine. The 2 atoms of O unite with each other, and thus neutralize 2 of their equivalents, leaving 2 unsaturated. Now 1 of these is neutralized by the Cl, leaving 1 unsaturated. Thus ClO_2 becomes a *uni*valent.

CHEMICAL TYPES.

7. *Types of Chemical Compounds.* — "There are three modes or forms of atomic grouping to which so large a number of substances may be referred, that they are regarded as molecular types or patterns, according to which the atoms of a molecule are grouped together. . . . . The three compounds, hydrochloric acid, water, and ammonia, $H\overset{I}{Cl}$, $H_2\overset{II}{O}$, and $H_3\overset{III}{N}$ are generally taken as the representatives of these types; and substances are said to belong to the type of hydrochloric acid, to

the type of water, or to the type of ammonia, as the case may be. . . . It will be noticed that in the first of these types a single univalent atom or radical is united to another single univalent atom; that in the second, a bivalent atom binds together two univalent atoms, or their equivalents; that in the third a trivalent atom binds together three univalent atoms, or their equivalents."

8. *Condensed Types.* — "In the same way that a bivalent atom may bind together two univalent atoms or their equivalents, so also it may serve to bind together two *molecules;* and, in like manner, a trivalent atom may bind together three *molecules* into a more complex molecular group; and thus are formed what are called *condensed types.*"

9. *Metallic Chlorides.* — All the metallic chlorides, fluorides, bromides, iodides, and cyanides may be considered as belonging to the muriatic acid type. In those of the monad metals, there is no condensation, or soldering together of molecules. Thus we have —

$$\text{HCl (muriatic acid).}$$
$$\text{AgCl (argentic chloride).}$$

Here one atom of hydrogen in a molecule of HCl may be regarded as replaced by an atom of a univalent metal.

The chlorides, etc., of the bivalent metals may be regarded as formed by the soldering together of 2 molecules of HCl by a bivalent atom which replaces the H. Thus, —

$$\left.\begin{matrix}HCl\\HCl\end{matrix}\right\}\text{ (two molecules of muriatic acid).}$$

$$Ca\left.\begin{matrix}Cl\\Cl\end{matrix}\right\} = CaCl_2\text{ (calcic chloride).}$$

The chlorides, etc., of the trivalent metals may be regarded as 3 molecules of HCl, soldered together by a trivalent atom, which has replaced the H. Thus, —

$$\left.\begin{matrix}HCl\\HCl\\HCl\end{matrix}\right\}\text{ (three molecules of muriatic acid).}$$

$$Au\left.\begin{matrix}Cl\\Cl\\Cl\end{matrix}\right\} = AuCl_3\text{ (auric chloride).}$$

The chlorides, etc., of the quadrivalent metals may be regarded as four molecules of HCl, soldered together by a quadrivalent atom which replaces the H. Thus, —

$$\left.\begin{matrix} HCl \\ HCl \\ HCl \\ HCl \end{matrix}\right\} \text{(four molecules of muriatic acid).}$$

$$Pt \left.\begin{matrix} Cl \\ Cl \\ Cl \\ Cl \end{matrix}\right\} = PtCl_4 \text{ (platinic chloride).}$$

The symbol of *aluminic chloride* is Al_2Cl_6. Here, although Al is *quadri*valent, Al_2 is only *sexti*valent, not *octi*valent. Two atoms of Cl seem to combine with each other, and thus neutralize 2 of their 8 equivalents. There are several cases like this. Chromium, iron, manganese, uranium, and some other metals are both bivalent and quadrivalent. When quadrivalent, they combine like aluminium. Thus we have —

Cr_2Cl_6, chromic chloride.
Fe_2Cl_6, ferric chloride.
Mn_2Cl_6, manganic chloride.
U_2Cl_6, uranic chloride.

As bivalents, they form the following chlorides: —

$CrCl_2$, chromous chloride.
$FeCl_2$, ferrous chloride.
$MnCl_2$, manganous chloride.
UCl_2, uranous chloride.

The bivalent *mercury* forms the two chlorides, $HgCl_2$, mercuric chloride, and Hg_2Cl_2, mercurous chloride. In the last, the 2 atoms of Hg combine with each other, and thus neutralize 2 of their 4 equivalents, leaving the other 2 to be neutralized by the Cl_2. The bivalent copper is like mercury in this respect.

10. *Bases.* — The bases belong to the *water* type, and may be regarded as made up by replacing a *part* of the H of the H_2O by an equivalent of a metallic element. In the bases of the *uni*valent metals, 1 atom of H is replaced by 1 of the univalent metal. Thus, —

$HHO = H_2O$.
KHO = potassic hydrate.

The bases of the *bi*valent metals are of a condensed type, 2 molecules of H_2O being soldered together by a bivalent atom, which replaces an atom of H in each. Thus, —

$H_2H_2O_2 = 2H_2O$
$CaH_2O_2 =$ calcic hydrate.

The bases of the *tri*valent metals are also condensed in type, 3 molecules of H_2O being soldered together by a trivalent atom. Thus, —

$H_3H_3O_3 = 3H_2O$.
$AuH_3O_3 =$ auric hydrate.

The quadrivalent *aluminium* forms a base by soldering together 6 molecules of H_2O with 2 trivalent atoms. Thus, —

$H_6H_6O_6 = 6H_2O$.
$Al_2H_6O_6 =$ aluminic hydrate.

It will be seen that this hydrate is analogous to the aluminic chloride already explained. In the same way, chromium, iron, manganese, and uranium, as quadrivalents, form the hydrates —

$Cr_2H_6O_6$, chromic hydrate.
$Fe_2H_6O_6$, ferric hydrate.
$Mn_2H_6O_6$, manganic hydrate.
$U_2H_6O_6$, uranic hydrate.

As bivalents, they form the bases —

CrH_2O_2, chromous hydrate.
FeH_2O_2, ferrous hydrate.
MnH_2O_2, manganous hydrate.
UH_2O_2, uranous hydrate.

Mercury also forms two bases, HgH_2O_2 and $Hg_2H_2O_2$, corresponding to its two chlorides.

11. *Basic Oxides.* — Basic oxides belong to the *water* type, and may be regarded as formed by replacing *all* the H of one or more molecules of H_2O by an equivalent of metal.

In the *uni*valent oxides, 2 equivalents of H are replaced by 2 univalent atoms. Thus, —

$$H_2O = \text{water.}$$
$$Ag_2O = \text{argentic oxide.}$$

In the *bi*valent oxides, 2 equivalents of H are replaced by 1 bivalent atom. Thus, —

$$H_2O = \text{water.}$$
$$CaO = \text{calcic oxide.}$$

In the *tri*valent oxides, 6 equivalents of H are replaced by 2 trivalent atoms, which bind together 3 molecules of H_2O. Thus, —

$$H_6O_3 = 3H_2O.$$
$$Au_2O_3 = \text{auric oxide.}$$

Al_2O_3, *aluminic oxide*, Fe_2O_3, *ferric oxide*, Cr_2O_3, *chromic oxide*, Mn_2O_3, *manganic oxide*, and U_2O_3, *uranic oxide*, are formed in the same way as the trivalent oxides; 2 of the metallic equivalents in each case being neutralized by the metal itself, as in their chlorides and hydrates. All these metals, except Al, form regular *bi*valent oxides, whose names take the ending *-ous*. Hg, besides its bivalent oxide, HgO, forms a *mercurous oxide*, Hg_2O.

12. *Acids.*—The oxyacids belong to the *water* type, and may be regarded as formed by replacing a *part* of the H of one or more molecules of H_2O by an equivalent of a non-metallic element or radical.

Nitrous acid is formed by replacing 3 atoms of H in 2 molecules of H_2O by 1 of the trivalent N. Thus, —

$$HH_3O_2 = 2H_2O.$$
$$HNO_2 = \text{nitrous acid.}$$

Nitric acid may be formed by replacing 1 atom of H in 1 molecule of H_2O by the univalent radical NO_2. Thus, —

$$HHO = H_2O.$$
$$HNO_2O = HNO_3 \text{ (nitric acid).}$$

In *chlorous* acid, 1 atom of H is replaced by the univalent radical ClO. Thus, —

$$HHO = H_2O.$$
$$HClOO = HClO_2.$$

If the atom of H be replaced by the univalent ClO_2, we get $HClO_3$, *chloric acid.*

When 2 molecules of water are soldered together by the bivalent O, we get H_2SOO_2 or H_2SO_3, *sulphurous acid.* If the bivalent SO_2 solders the 2 molecules of H_2O, we get $H_2SO_2O_2$ or H_2SO_4, *sulphuric acid.*

When 3 molecules of H_2O are soldered by the trivalent P or As, we get H_3PO_3, *phosphorous acid,* and H_3AsO_3, *arsenious acid.* If the 3 molecules of H_2O are soldered by the trivalent PO or AsO, we get H_3PO_4, *phosphoric acid,* or H_3AsO_4, *arsenic acid.*

Boracic acid, HBO_2, is formed by replacing 3 atoms of H in 2 molecules of H_2O by the trivalent B.

Silicic acid, H_4SiO_4, is formed by replacing 4 atoms of H in 4 molecules of H_2O by the quadrivalent Si.

Carbonic acid, H_2CO_3, is formed by replacing 2 atoms of H in 2 molecules of H_2O by the bivalent CO.

13. *Anhydrides.* — The *anhydrides* belong to the *water* type, and are formed by replacing *all* the H of one or more molecules of H_2O by an equivalent of a non-metallic element or radical. Thus, —

$$HHO = H_2O.$$
$$\overset{\mathrm{I}}{NO_2}\,\overset{\mathrm{I}}{NO_2}O = N_2O_5 \text{ (nitric anhydride).}$$
$$H_2O = \text{water.}$$
$$\overset{\mathrm{II}}{C}OO = CO_2 \text{ (carbonic anhydride).}$$
$$\overset{\mathrm{II}}{SO_2}O = SO_3 \text{ (sulphuric anhydride).}$$
$$H_3H_3O_3 = 3H_2O.$$
$$\overset{\mathrm{III}}{PO}\,\overset{\mathrm{III}}{PO}O_3 = P_2O_5 \text{ (phosphoric anhydride).}$$

14. *Salts.* — When the H of a base is replaced by a non-metallic element or radical, or the H of an acid by a metallic element or radical, a *salt* is formed. Thus, —

KHO gives $KClO$ (potassic hypochlorite).
$NaHO$ gives $NaNO_2O = NaNO_3$ (sodic nitrate).

CaH_2O_2 gives $CaSO_2O_2 = CaSO_4$ (calcic sulphate).
CaH_2O_2 gives $CaCOO_2 = CaCO_3$ (calcic carbonate).

Again, —

HNO_3 gives KNO_3 (potassic nitrate).
H_2SO_4 gives $CuSO_4$ (cupric sulphate).
H_3PO_4 gives K_3PO_4 (potassic phosphate).

The *nitrates* of the metals are analogous to the *chlorides*, with NO_3 in place of the Cl. The *sulphates* are analogous to the *oxides*, with SO_4 in place of the O. Thus, —

KCl	KNO_3 (potassic nitrate).
$CaCl_2$	$Ca2NO_3$ (calcic nitrate).
Al_2Cl_6	Al_26NO_3 (aluminic nitrate).
K_2O_2	K_2SO_4 (potassic sulphate).
CaO	$CaSO_4$ (calcic sulphate).
Al_2O_3	Al_23SO_4 (aluminic sulphate).

The *chlorates* are like the *nitrates;* and the *carbonates* like the *sulphates*, with CO_3 in place of SO_4.

The *sulphides* are like the *oxides*.

15. *Substitution.* — "When cotton-wool is dipped in strong nitric acid (rendered still more active by being mixed with twice its volume of concentrated sulphuric acid), and afterwards washed and dried, it is rendered highly explosive; and, although no important change has taken place in its outward aspect, it is found on analysis to have lost a certain amount of hydrogen, and to have gained from the nitric acid an equivalent amount of nitric peroxide, NO_2, in its place."

$C_{12}H_{20}O_{10}$ (cotton) becomes $C_{12}H_{14}(NO_2)_6O_{10}$ (gun-cotton).

Under the same conditions, glycerine undergoes a like change, and is converted into the explosive nitro-glycerine.

$C_6H_{16}O_6$ (glycerine) becomes $C_6H_{10}(NO_2)_6O_6$ (nitro-glycerine).

So also the hydro-carbon, naphtha, called *benzole*, is changed into nitro-benzole.

C_6H_6 (benzole) becomes $C_6H_5(NO_2)$ (nitro-benzole).

"The last compound is not explosive, and the explosive nature of the first two is, in a measure, an accidental quality, and is evidently owing to the fact, that into an already complex structure there have been introduced, in place of the indivisible atoms of hydrogen, the atoms of a highly unstable radical, rich in oxygen. The point of chief interest for our chemical theory is that this substitution does not alter, at least profoundly, the outward aspect of the original compound. Every one knows how closely gun-cotton resembles cotton-wool. In like manner, nitro-glycerine is an oily liquid, like glycerine; and nitro-benzole, although darker in color, is a highly aromatic, volatile fluid, like benzole itself. Products like these are called *substitution products;* and they certainly suggest the idea that each chemical compound has a certain definite structure, which may be preserved even when the materials of which it is built are, in part, at least, changed. If, in the place of firm iron girders, we insert weak wooden beams, a building, while retaining all its outward aspects, may be rendered wholly insecure. And so the explosive nature of the products we have been considering is not at all incompatible with a close resemblance, in outward aspects and internal structure, to the compounds from which they are derived."

16. *Isomorphism.* — "Closely associated with the facts of the last section, which find their chief manifestation in substances of organic origin, are the phenomena of *isomorphism*, which are equally conspicuous among artificial salts and native minerals. There seems to be an intimate connection between chemical composition and crystalline form, and two substances which, under a like form, have an analogous composition, are said to be *isomorphous*." The following minerals are isomorphous: —

Calcite, or calcic carbonate,	$CaCO_3$
Magnesite, or magnesic carbonate,	$MgCO_3$
Chalybdite, or ferrous carbonate,	$FeCO_3$
Diallogite, or manganous carbonate,	$MnCO_3$

"The most cursory examination of these symbols will show that they differ from each other only in the fact that one metallic atom has been replaced by another. It is not, however, every metallic atom which can thus be put in without altering the form. This is a peculiarity which is confined to certain groups of elements, which for this reason are called *groups of isomorphous elements*. Moreover, as a rule, there is a close resemblance between the members of any one of these groups in all their other chemical relations. These facts, like those of the last section, tend to show that the molecules of every substance have a determinate structure, which admits of a limited substitution of parts without undergoing essential change, but which is either destroyed or takes a new shape when, in place of one of its constituents, we force in an unconformable element. A well-known class of artificial salts called the *alums* affords even a more striking illustration of the principles of isomorphism than the simpler examples we have chosen."

17. *Isomerism, Allotropism, and Polymorphism.* — "We should infer from the doctrine of chemical types that the same atoms might be grouped together in different ways, so as to form different molecules, which, in their aggregation, would present essentially distinct qualities. Hence we should expect to find distinct substances having the same composition, and, in fact, our science, organic chemistry especially, is rich in examples of this kind. Such substances are said to be *isomeric*, and the phenomenon is called *isomerism*." Thus, common *sugar* and *gum arabic* have exactly the same composition, $C_{12}H_{22}O_{11}$, and are *isomeric*.

When, however, the differences are not sufficiently great to justify a distinct name, the two bodies are said to be different *allotropic* states of the same substance, and the phenomenon is called *allotropism*. Thus, there are three varieties of *tartaric acid* which differ in their action on polarized light, but which are in almost every other respect identical.

Sometimes a substance *crystallizes* in fundamentally different forms, as calcic carbonate in the minerals *calcite* and *aragonite*. This phenomenon is called *polymorphism*, and is invariably accompanied by a marked difference of properties.

Differences of this kind, even more marked, are found among elementary substances, and give strong grounds for believing that these elements may be really composite.

ELECTRICAL RELATIONS OF THE ELEMENTS.

18. When the electric current is passed through any compound liquid which is a conductor of electricity, the liquid is decomposed; its atoms, or radicals, travelling in opposite directions towards the two poles. Thus, when the current is sent through muriatic acid, the hydrogen atoms go towards the negative pole, or *cathode*, and the chlorine atoms towards the positive pole, or *anode*. There are certain elements and radicals which, like hydrogen, always go towards the negative pole; and certain others, like chlorine, which always go towards the positive pole. The former are called *electro-positive*, and the latter *electro-negative* elements. There are other elements which in some of their compounds are electro-positive, while in others they are electro-negative. The electrical relations of the elements are best shown by grouping the elements in series, so that each member of the series shall be electro-positive when in combination with the elements which follow it, and electro-negative when combined with those which precede it. It is found that the non-metals are always negative when combined with the metals, which are then positive. The electrical relations of the metals show that *hydrogen*, although a gas, and the lightest substance known, is really a *metal*. It is the one gaseous metal, as mercury is the one liquid metal.

CLASSIFICATION OF THE ELEMENTS.

19. *The Metals.* — The electrical relations of the elements, as well as their formation of bases and acids, show that there is a fundamental distinction between metals and non-metals. "The characteristic qualities of a metal, with which every one is more or less familiar, are the so-called metallic lustre, that peculiar adaptability of molecular structure known as malleability or ductility, and the power of conducting electricity or heat. These qualities are found united and in their perfection only in the true

metals, although one, or even two of them, are well developed in several elementary substances, which, on account of their chemical qualities, are now almost invariably classed among the non-metals; as, for example, selenium, tellurium, arsenic, antimony, boron, and silicon."

20. *Scheme of Classification.* — "The classification of the elementary atoms which has been adopted in this book is shown in the table on page 180. The atoms are classed, in the first place, in six families, according as in their usual condition they are equivalent to one, two, three, four, five, or six atoms of hydrogen. These families are subdivided into groups, separated by bars in the table, of closely allied elements. The atoms of any one of these groups are *isomorphous*, and they are arranged in the order of their weights, which is found to correspond also, in almost every case, with their electrical relations. Each group forms a very limited chemical series; and not only the weights and the electrical relations of the atoms, but also many of the physical qualities of the elementary substances vary regularly as we pass from one end of the series to the other. The order of the variation, however, is not always the same; for while, in some cases, the lightest atoms of a series are the most electro-negative, in other cases they are the most electro-positive."

DOUBLE DECOMPOSITION.

21. The salts, both binary and ternary, are ordinarily formed by a double transfer of elements, as shown in the following equation: —

$$\overset{\text{I}}{H}_2SO_4 + \overset{\text{II}}{Ba}H_2O_2 = \overset{\text{II}}{Ba}SO_4 + 2H_2O.$$

Such a double transfer of elements is called *double decomposition*. The change which takes place is usually called simply a *reaction*.

On mixing any two solutions, such a change will always take place, when an *insoluble* or *volatile* compound would thereby be formed.

In working out the following reactions, it must be borne in mind that it is always the *metals* which change places; and that they must always be in equivalent quantities. It is better in all cases to indicate by the proper mark the combining power of the elements which are to replace each other. Before beginning to work the problems of the next section, the student should be able to read and explain all the equations given below, and especially to give the reason for the number of molecules used in each case. If more molecules than one are used in any case, it is always in order that the combining units of the elements which replace each other may be equal.

$$\overset{I}{H}_2S + \overset{II}{Pb}2NO_3 = 2\overset{I}{H}NO_3 + \overset{II}{Pb}S.$$
$$2\overset{I}{H}NO_3 + \overset{II}{Ba}H_2O_2 = \overset{II}{Ba}2NO_3 + 2\overset{I}{H}_2O.$$
$$3\overset{I}{H}_2S + 2\overset{III}{As}Cl_3 = 6\overset{I}{H}Cl + \overset{III}{As}_2S_3.$$
$$\overset{III}{Bi}Cl_3 + 3\overset{I}{K}I = \overset{III}{Bi}I_3 + 3\overset{I}{K}Cl.$$
$$2\overset{I}{H}Cl + \overset{II}{Pb}2NO_3 = 2\overset{I}{H}NO_3 + \overset{II}{Pb}Cl_2.$$
$$2\overset{I}{Ag}NO_3 + \overset{II}{Ca}Cl_2 = \overset{II}{Ca}2NO_3 + 2\overset{I}{Ag}Cl.$$
$$\overset{II}{Ca}SO_4 + \overset{II}{Ba}2NO_3 = \overset{II}{Ca}2NO_3 + \overset{II}{Ba}SO_4.$$

PROBLEMS.

1. Show the reaction between plumbic nitrate and sodic hydrate.

$$\overset{II}{Pb}2NO_3 + 2H\overset{I}{Na}O = \overset{II}{Pb}H_2O_2 + 2\overset{I}{Na}NO_3;$$

that is, plumbic nitrate *plus* sodic hydrate gives plumbic hydrate *plus* sodic nitrate.

2. Plumbic nitrate and sulphuric acid.
3. Cupric sulphate and potassic hydrate.
4. Manganic sulphate and sodic hydrate.
5. Zincic sulphate and potassic hydrate.
6. Cobaltic nitrate and sodic hydrate.
7. Manganic sulphate and ammonic carbonate, $(H_4N)_2CO_3$.
8. Baric nitrate and ammonic carbonate.
9. Calcic sulphate and ammonic carbonate.
10. Strontic nitrate and calcic sulphate.

11. Aluminic sulphate and ammonic hydrate.
12. Ferrous nitrate and ammonic hydrate.
13. Hydric chloride and argentic nitrate.
14. Hydric chloride and plumbic nitrate.
15. Hydric chloride and mercurous nitrate.
16. Hydric sulphide and argentic nitrate.
17. Hydric sulphide and plumbic nitrate.
18. Hydric sulphide and mercurous nitrate.
19. Argentic nitrate and potassic iodide.
20. Potassic iodide and plumbic nitrate.
21. Potassic iodide and mercurous nitrate.
22. Stannous chloride ($SnCl_2$) and hydric sulphide.
23. Stannic chloride ($SnCl_4$) and hydric sulphide.
24. Bismuthic chloride and hydric sulphide.
25. Stannous chloride and potassic hydrate.
26. Auric chloride and hydric sulphide.
27. Mercuric chloride and hydric sulphide.
28. Platinic chloride and potassic iodide.
29. Auric chloride and potassic iodide.
30. Cupric sulphate and hydric sulphide.
31. Mercuric chloride and potassic iodide.
32. Cupric sulphate and potassic iodide.
33. Aluminic chloride and ammonic hydrate.
34. Baric chloride and ammonic carbonate.
35. Calcic chloride and ammonic carbonate.
36. Baric chloride and sodic hydrate.
37. Strontic chloride and calcic sulphate.
38. Argentic nitrate and ammonic hydrate.
39. Argentic nitrate and sulphuric acid.
40. Cadmic chloride and hydric sulphide.

If additional problems are desired, they can be easily prepared by the teacher.

The Table on the following page is from Professor Cooke's "First Principles of Chemical Philosophy." All the direct quotations in this chapter are from the same book, our great indebtedness to which has already been acknowledged in the Preface.

TABLE.

UNIVALENTS.

Metals.	Atomic Weights.	Symbols of Molecules.	Quantivalence.
Hydrogen	1.0	*H-H*	I
Lithium	7.0	*Li-Li*	,,
Sodium	23.0	*Na-Na*	,,
Potassium	39.1	*K-K*	,,
Rubidium	85.4	*Rb-Rb*	,,
Cæsium	133.0	*Cs-Cs*	,,
Silver	108.0	*Ag-Ag*	,,
Thalium	204.0	*Tl-Tl*	I or III
Non-Metals.			
Fluorine	19.0	*F-F*	I
Chlorine	35.5	*Cl-Cl*	,,
Bromine	80.0	*Br-Br*	,,
Iodine	127.0	*I-I*	,,

BIVALENTS.

Metals.	Atomic Weights.	Symbols of Molecules.	Quantivalence.
Calcium	40.0	*Ca*	II
Strontium	87.6	*Sr*	,,
Barium	137.0	*Ba*	,,
Lead	207.0	*Pb*	II or IV
Magnesium	24.0	*Mg*	II
Zinc	65.2	*Zn*	,,
Indium	72.0	*In*	,,
Cadmium	112.0	*Cd*	,,
Glucinum	9.3	*G ?*	,,
Ytrium	61.7	*Y ?*	,,
Erbium	112.6	*E ?*	,,
Lanthanum	93.6	*La ?*	
Didymium	95.0	*D ?*	
Manganese	55.0	*Mn*	II or IV
Iron	56.0	*Fe*	,,
Nickel	58.8	*Ni*	,,
Cobalt	58.8	*Co*	,,
Copper	63.4	*Cu*	II
Mercury	200.0	*Hg*	,,
Non-Metals.			
Oxygen	16.0	*O*	,,
Sulphur	32.0	*S*	II or VI
Selenium	79.4	*Se*	,,
Tellurium	128.0	*Te*	,,

TRIVALENTS.

Metals.	Atomic Weights.	Symbols of Molecules.	Quantivalence.
Gold	197.0	*Au-Au*	III
Non-Metals.			
Boron	11.0	*B ?*	,,
Nitrogen	14.0	*N-N*	III or V
Phosphorus	31.0	P_2-P_2	,,
Arsenic	75.0	As_2-As_2	,,
Antimony	122.0	Sb_2-Sb_2	,,
Bismuth	210.0	Bi_2-Bi_2	,,

QUADRIVALENTS.

Metals.	Atomic Weights.	Symbols of Molecules.	Quantivalence.
Aluminium	27.4	*Al ?*	IV
Chromium	52.2	*Cr ?*	II or IV
Cerium	92.0	*Ce ?*	,,
Uranium	120.0	*U ?*	,,
Ruthenium	104.4	*Ru ?*	,,
Osmium	199.2	*Os ?*	,,
Rhodium	104.4	*R ?*	,,
Iridium	196.0	*Ir ?*	,,
Palladium	106.6	*Pd ?*	,,
Platinum	197.4	*Pt ?*	,,
Titanium	50.0	*Ti ?*	IV
Tin	118.0	*Sn ?*	,,
Zirconium	89.6	*Zr ?*	,,
Thorium	231.5	*Th ?*	,,
Non-Metals.			
Carbon	12.0	*C ?*	IV
Silicon	28.0	*Si ?*	,,

QUINTIVALENTS.

Metals.	Atomic Weights.	Symbols of Molecules.	Quantivalence.
Niobium	94.0	*Nb ?*	V
Tantalum	182.0	*Ta ?*	,,

SEXTIVALENTS.

Metals.	Atomic Weights.	Symbols of Molecules.	Quantivalence.
Molybdenum	96.0	*Mo ?*	VI
Tungsten	184.0	*W ?*	,,
Vanadium	137.0	*V ?*	,,

NOTES ON EXPERIMENTS.

APPARATUS.

The following list includes all the apparatus needed for the more important experiments in this book: —

1. Pneumatic trough. — Gases which are not absorbed by water, or but slightly so, can be collected by first filling the vessel with water, and inverting it with its mouth under water. The tube through which the gas is escaping is introduced under the mouth of the jar, and the gas rises and fills the jar.

For collecting gases in this way, a pneumatic trough is useful and almost indispensable. This apparatus consists of a vessel deep enough to allow of filling and inverting under the water any of the jars ordinarily used. A shelf perforated with holes extends across the vessel about an inch under the surface of the water. After the jar is filled with water and inverted, it is set upon the shelf over one of the holes. The delivery tube (which, except the part passing through the cork, may be of rubber) is then passed into this hole, and the jar filled. It is convenient to have a shelf large enough to hold several jars, after they have been filled with gas.

The best pneumatic trough is made of soapstone or copper; but any vessel which will hold water and has a perforated shelf will answer the purpose. The teacher can easily provide himself with such a trough at little expense.

2. Two glass cylinders: one, 1 inch in diameter and 5 inches deep; the other, 1½ inches in diameter and 7 inches deep. These should have ground mouths and be provided with ground-glass plates for covers.

3. Two two-quart glass jars, with ground mouths and ground plates for covers.

4. Two nipper-taps. A nipper-tap is a convenient substitute for a stop-cock. By means of it a rubber tube can be pinched close. The best kind is shown (full size) in Figure 32. The tube is compressed by a spring.

Fig. 32.

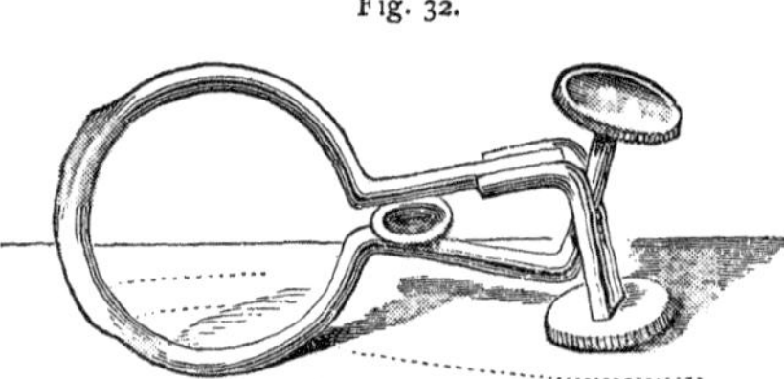

5. One two-gallon india-rubber gas-bag. This may be closed with a rubber cork, through which passes a short glass tube, connected with a second glass tube by a short piece of rubber tubing, which may be closed with a nipper-tap. Such an arrangement is cheaper than a stop-cock, and better, since gases sometimes corrode the latter.

6. A bottle generator. This consists of a wide-mouthed pint bottle, closed with a rubber cork, through which pass two tubes; one, large and straight, and passing nearly to the bottom; the other, small, and bent at right angles, passing only through the cork. An ordinary cork will answer, but rubber corks are always to be preferred for chemical apparatus. They cost more at first, but they last much longer, and make perfectly tight joints.

7. Three half-pint flasks. One, at least, of these should have a rubber cork through which passes a bent delivery-tube.

8. An iron lamp-stand with three rings.

9. A lamp. A Bunsen's lamp is much to be preferred where gas is to be had; otherwise, an alcohol lamp.

10. A wooden retort-holder with two clamps.

11. A hydrogen generator. The most convenient apparatus for preparing hydrogen is the self-regulating generator shown in Figure 33. It consists of a glass vessel closed with a metallic cap. A bell-shaped glass vessel, open at top and bottom, is fastened to this cap by an air-tight joint. A tube closed by a stop-cock passes through the cap into the bell-shaped vessel. A copper bucket perforated with fine holes is hung from a hook

Fig. 33.

inside this vessel. This bucket is filled with shreds of sheet zinc or bits of granulated zinc. The outer vessel is filled about two-thirds full of dilute sulphuric acid (the ordinary oil of vitriol diluted with about ten parts of water). The metallic cap with the bell-glass attached is then put in its place, and the stop-cock opened. The air is first driven out, and the dilute acid coming in contact with the zinc begins to act upon it, and hydrogen is given off in abundance. When sufficient hydrogen has been obtained, the stop-cock is closed, the hydrogen generated collects in the upper part of the bell-glass, and drives the acid out. As soon as the acid is driven out, the action ceases until the stop-cock is opened again, and the hydrogen allowed to escape. By means of rubber tubing, the hydrogen can be conveyed from the generator to the jar or vessel in which it is to be collected. When hydrogen is to be burned, the greatest care must be taken that all the air is driven out of the apparatus before the gas is collected.

12. A chlorine tube. This is a glass-tube, ¾ of an inch in diameter, and 15 inches long, closed at one end. The other end should be provided with a rubber cork, through which passes a glass tube drawn out to a fine jet inside. This should be connected with a small glass funnel by means of a short piece of rubber tubing, which may be closed by a nipper-tap.

13. Two small beakers.

14. Two evaporating dishes.

15. A U-tube (Figure 5). One end of the tube should be closed, and near this closed end two platinum wires should pass through the sides so as nearly to meet within. These wires should terminate outside in loops. It is not necessary that the U-tube should be tubulated at the bend, as represented in the figure, or fastened to a stand, since it may be held in the clamp of the retort-holder.

16. A chalk cup. This is a cylinder of chalk, about 1½ inches in diameter and 2 inches high, hollowed out at the top.

17. A wash-bottle. This is like No. 6. The rubber tube which

is delivering the gas should be connected with a glass tube small enough to pass down through the straight glass tube to the bottom.

18. A nitric-oxide bell and jar. The bell is half the capacity of the jar, and the mouths of both are ground, so as to fit together air-tight.

19. An asbestos tube.

20. A half-pint tubulated retort.

21. A dozen 5-inch test-tubes.

22. Half a pound of glass tubing, ¼ of an inch in diameter.

23. 4 feet of rubber tubing, ¼ of an inch in diameter.

The following additional apparatus will be needed for experiments with the oxy-hydrogen blowpipe: —

Two 25-gallon gas-bags, with stop-cocks and tubing.

An oxy-hydrogen jet.

A copper flask for making oxygen.

EXPERIMENTS.

1. (§ 1, *Exp.* 1.) Use bottle generator (No. 6 in list of apparatus above). The end of the long tube must be kept beneath the surface of the acid. The acid is diluted with 2 or 3 parts of water. The hydrogen may be collected in the small glass cylinder (No. 2); but the gas which first comes off must not be collected, as it is mixed with air, and therefore explosive.

2. (§ 1, *Exp.* 2.) Use flask (No. 7) with rubber cork. Pour in strong muriatic acid first, so that none of the dry manganese may stick to the flask. Use only a small quantity of the materials, as the chlorine is very disagreeable if it escapes into the room. It is well to have a large jar filled with water, ready to receive any chlorine that may come off after the small cylinder has been filled. It will be well also to let the gas which first comes off pass into this jar until the air is expelled from the flask. The flask should be supported on the ring of the lampstand (No. 8), and should be separated from the flame, either by a sand-bath or, better, by a piece of fine wire gauze.

3. (§ 1, *Exp.* 3.) Use the smaller glass cylinder (No. 2). Let the hydrogen pass in first until it is half full, and then fill

the remainder with chlorine. Close the cylinder under water with the glass plate, bring it near the lamp, remove the glass plate, and quickly put the mouth of the cylinder into the flame. The explosion is not very violent with these small quantities. Never attempt to use large quantities, and be careful to keep the mixed gases out of direct sunshine.

Fig. 34.

Another way of performing the experiment is to use two cylinders of the smaller size (No. 2), filling one with hydrogen and the other with chlorine. Cover them with the glass plates, and place them together, mouth to mouth, as represented in Figure 34. Then withdraw the glass plates, keeping the mouths of the cylinders together, shake the cylinders to mix the gases, and open them over a burning lamp, as shown in Figure 35.

Fig. 35.

4. (§ 2, *Exp.* 1.) The sodium used must be pure, else there is danger of an explosion. The smaller cylinder (No. 2) must be used and held in the clamp of the retort-holder (No. 10). The sodium may be first thrown upon the water, and then thrust under the cylinder by means of a perforated metallic cup, which may be made out of a thimble by punching a few holes through the top, and fastening a wire round it for a handle.

5. (§ 2, *Exp.* 2.) The water may be saturated with chlorine by putting it in a wash-bottle (No. 17), and allowing the chlorine to pass through it as long as it is taken up by the water.

The delivery-tube of the bottle should be connected with a jar over the water-trough, in order that no chlorine may escape into the room. The chlorine water thus prepared may be put in the larger cylinder (No. 2), inverted in a soup-plate of water, and set in the sunshine.

6. (§ 2, *Exp.* 3.) Fill one of the jars (No. 3) with oxygen, and let it stand over the trough long enough to drain. Burn the hydrogen as it escapes from the generator (No. 11) through a glass tube bent at right angles and drawn out to a fine jet; and hold the jar of oxygen over the flame.

Glass tubing may be bent by heating it red-hot in the flame of the lamp. A tube may be drawn out to a fine jet by heating the middle red-hot, and then quickly pulling the ends apart. It may be cut by scratching it a little on one side with a three-cornered file, and then bending it over the thumb-nails, held close together on the side opposite the scratch.

7. (§ 3, *Exp.* 1.) Fill chlorine tube (No. 12) with chlorine over the water-trough. Close the mouth while under water with the thumb, invert it, and insert the cork. Pour a little strong ammonia into the funnel, and let it drop slowly into the tube by means of the nipper-tap, to the depth of half an inch. Close the nipper-tap, and pour off the ammonia from the funnel; and then fill the funnel with water, and let it run into the tube, care being taken to keep water in the funnel all the time, so that no air may get into the tube.

In this and all other cases where chlorine is to be used, it will be best to prepare it beforehand, and keep it in a gas-bag till wanted. On its way from the generating flask to the gas-bag, it should be passed through a wash-bottle containing a small amount of water. The gas is washed to remove any muriatic acid which may be in it, and which would corrode the bag.

8. (§ 6, *Exp.* 1.) Fill the bent tube in the first place with water, close it with the thumb, and invert it over water. Then allow a little hydrogen to pass into it from the generator, and also some oxygen from a gas-bag which has been previously filled with this gas. With a little care, one can readily get the right proportions. The tube should not be more than two-thirds full of the mixed gases, and a considerable excess of one

gas should be used, so that enough may be left for testing. After putting the gases into the tube, close it with the thumb, raise it from the water, and tip it so that the gases may pass into the closed arm; then pour out a little of the water, and again close it with the thumb. Connect the outer coating of a charged Leyden jar with one of the platinum loops, by means of a wire or chain, and bring the knob of the jar in contact with the other loop. After the spark has passed, again fill the open arm with water, close it with the thumb, and tip the tube so as to transfer the remaining gas to the open arm, and there test it.

9. (§ 10, *Exp.* 1.) As explosions sometimes occur in the manufacture of oxygen, care should be taken that the chemicals are pure; and it would be well to test each purchase by putting some of the potassic chlorate in an iron spoon and heating it over a spirit lamp until it is melted; then stir into it with an iron wire some of the black oxide of manganese; and if these materials are not good, an explosion will take place, and a whitish mass with red spots in it will be left in the spoon. If, however, the chemicals are pure, there will be no explosion, and the melted mixture will soon dry up, leaving a dark gray residuum. If the bubbles come over too violently, remove the lamp for a few minutes until they come more moderately.

A glass flask may be used; but as it is liable to break, a copper flask may perhaps be found cheaper in the long run, especially if one has to make oxygen often.

10. (§ 11, *Exp.* 1.) Use glass jar (No. 3). The charcoal may be fastened to a piece of wire and ignited in the flame of a lamp. It is well to have the other end of the wire attached to a copper or wooden disk, which serves to cover the jar while the charcoal is burning.

Lime-water may be made by putting a small quantity of slaked lime into a jar (No. 3) of water, stirring it thoroughly, and allowing it to settle. The clear liquid may then be poured off into a bottle, and kept for use.

11. (§ 11, *Exp.* 2.) Fill a jar (No. 3) with oxygen, and let it stand to drain. Put a piece of sulphur in the chalk cup (No. 16), ignite it by touching it with a red-hot wire, and invert a jar of oxygen quickly over it. The jar should be kept closed

with the plate until it is inverted, and brought just over the sulphur. The chalk cup should stand in a shallow dish of water, so that the mouth of the jar may dip under the water and prevent the SO_2 from escaping into the room.

12. (§ 11, *Exp.* 3.) The phosphorus may be burnt in the same way as the sulphur. As phosphorus is a very inflammable substance, and as burns from it are very slow to heal, great caution must be exercised in using it. The sticks must always be cut under water, and it must be carefully dried, by pressing it between pieces of unsized paper, before it is burnt; else it will fly about in a very disagreeable and dangerous manner. It should always be lighted by touching it with a heated wire, never by holding it in the flame of a lamp.

13. (§ 11, *Exp.* 4.) The watch-spring should be either straightened out or formed into a spiral, and fastened to a disk of copper or wood, as above described. Water should be left in the bottom of the jar to the depth of two inches or so, that the melted globules of iron may not burn into the jar.

14. (§ 12, *Exp.* 1.) Ozone may be most easily prepared by pouring a little ether into a jar (No. 3), and then when the air in the jar has become saturated with the vapor of ether, stirring the air with a glass rod heated nearly to redness.

15. (§ 20, *Exp.* 1.) In making hydrogen, use the hydrogen generator (No. 11).

16. (§ 21, *Exp.* 1.) Mix the gases in the gas-bag, letting each gas pass in separately. The quantities can be estimated nearly enough by the eye. Fasten an ordinary clay tobacco-pipe to the gas-bag, hold the mouth of the pipe in soap-suds, and press out the mixed gases in a gentle stream. For obvious reasons, the nipper-tap should be closed before the bubbles are lighted.

17. (§ 28, *Exp.* 1.) Fill the bottle generator (No. 6) one-third full with strong ammonia, and connect the large tube with a rubber bag filled with chlorine. Connect the delivery-tube with the wash-bottle (No. 17), and the delivery-tube of the wash-bottle with the larger cylinder (No. 2). Force the chlorine through the apparatus by compressing the bag. Care must be taken not to force an excess of chlorine through the ammonia,

lest the very explosive *nitric chloride* (page 28) be formed. There is little danger, however, if *strong* ammonia is used.

18. (§ 30, *Exp.* 1.) Use the bottle generator (No. 6). First pour in water enough to cover the bits of copper (brass or nickel will do as well), and then pour in gradually strong nitric acid. The gas is very poisonous, and should not be allowed to escape into the room.

19. (§ 30, *Exp.* 2.) The management of the experiment is the same as in the burning of phosphorus in oxygen (12).

20. (§ 30, *Exp.* 3.) The gases may be mixed in a gas-bag. It is not necessary that the proportions should be exact.

21. (§ 33, *Exp.* 1.) Use nitric-oxide bell and jar (No. 18). Fill the bell with oxygen, and the jar with nitric oxide. The jar should be closed with a plate and set on the table, and the bell of oxygen set on the plate covering the jar (Figure 36). When the experiment is to be tried, the plate should be drawn out, and the mouths of the bell and jar allowed to come together.

Fig. 36.

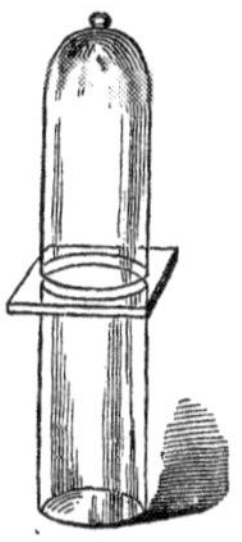

22. (§ 36, *Exp.* 1.) Use tubulated retort (No. 20). Only a small quantity of the ammonic nitrate need be used. The retort must not be heated too much, lest the gas should come off with explosive violence.

23. (§ 37, *Exp.* 1.) Gases, like ammonia and muriatic acid, which are absorbed by water, must be collected over mercury.

For the experiments of this book we think a mercury-trough a needless luxury. An evaporating dish of Berlin porcelain answers every purpose. One five inches in diameter is large enough to use with the cylinders mentioned above. It should be set in a shallow dish of strong glass or earthen ware, to prevent loss of mercury in case of accident. It should then be filled nearly full of mercury. The cylinder to be used is next filled with mercury to the brim. The ground-glass cover is then placed on the mouth of the cylinder and held firmly with the right hand, the cylinder inverted, and its mouth plunged beneath the mercury in the evaporating dish. It is well to put the cylinder in a strong dish when you are filling and inverting it, to save any mercury which may be spilled. After the cylinder is filled and inverted over the mercury in the dish, the plate is slipped from its mouth, and the cylinder is securely fastened in the clamp of a retort-stand, so that its mouth is held about ½ or ¾ of an inch above the bottom of the dish.

To fill the cylinder with ammonia gas, a little aqua ammonia is put into a flask provided with a delivery-tube, and gently boiled. To the delivery-tube is attached a piece of rubber tubing to connect it with the cylinder. A short piece of glass tubing with its end slightly turned up must be fastened to the end of the rubber tube, so that the gas may be introduced into the cylinder through the mercury. On first boiling the aqua ammonia, a large amount of air passes over from the flask. The ammonia gas must not be collected until this has all passed over. To determine when this has taken place, put the glass tube at the end of the rubber tube into a vessel of water; when heat is first applied to the flask, bubbles of gas will rise from the end of the tube through the water as long as any air comes over. When the air is all over, the ammonia gas is absorbed by the water as fast as it comes over, and no bubbles rise to the surface. After the air ceases to come over, plunge the end of the glass tube into the mercury, and introduce it under the mouth of the cylinder. The gas will rise into the jar and displace the mercury.

Fig. 37.

When the cylinder is filled with ammonia, close it with a glass plate, and remove it to a vessel of water; then open the mouth of the jar under the water (Figure 37).

24. (§ 43, *Exp.* 1.) The oxalic acid may be heated with sulphuric acid in a flask, and the solution of soda put in a wash-bottle.

25. (§ 45, *Exp.* 1.) In both experiments, the sulphur should be heated cautiously, that it may not

take fire. Figure 38 shows the needle-shaped crystals which may be obtained by the first experiment.

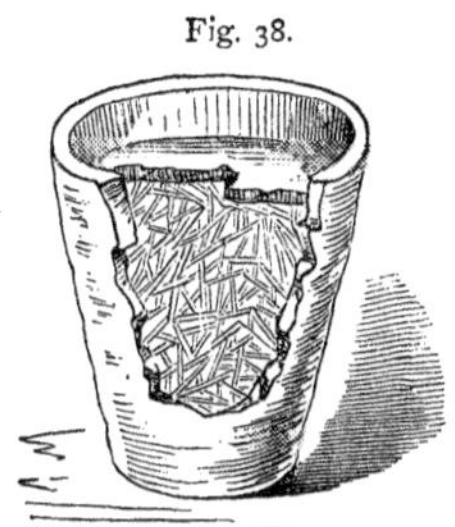
Fig. 38.

26. (§ 53, *Exp.* 1.) With the aid of an air-pump, phosphorus may be burnt in chlorine in the following manner: Place the chalk cup, with a bit of phosphorus in it, on the bottom of the nitric-oxide jar (No. 18), and cover the jar with a well-fitting glass plate which has in its centre a round hole, ¾ of an inch in diameter. This hole is closed with a rubber cork, through which passes a glass tube connected with a second glass tube by a short piece of rubber tubing provided with a nipper-tap. Connect the jar with the air-pump by rubber tubing, and exhaust the air. Close the nipper-tap, and connect the jar with a gas-bag filled with chlorine. Open the nipper-taps, and let the chlorine pass into the jar, until the pressure within is nearly but not quite equal to that outside. The phosphorus will soon take fire. If the jar were filled full of chlorine, the heat of the burning phosphorus would expand the chlorine, and force out some of it into the air.

27. (§ 60.) The color of iodine vapor may be shown by heating a few grains of iodine in a flask over the lamp.

28. (§ 65, *Exp.* 1.) Use tubulated retort (No. 20). Fill it nearly full of water, put in a small piece of phosphorus, and let the mouth of the retort dip under water. Heat the retort until the water in it boils, and then drop in a stick of caustic potash through the tubulature. The water is boiled first in order to drive out the air from the retort, so that the hydric phosphide may not take fire inside, and thus cause an explosion.

29. (§ 123, *Exp.* 1.) The process is the same as in the experiment in § 33, using air instead of oxygen.

30. (§ 123, *Exp.* 2.) Observe the precautions given in § 12.

31. (§ 142, *Exp.* 1.) The sugar and the potassic chlorate should be pulverized separately, and then mixed carefully. Put the mixture on an earthen dish, and let one drop of sulphuric acid fall upon it from the end of a long glass rod. The experiment is brilliant, even with much smaller quantities.

QUESTIONS FOR REVIEW AND EXAMINATION.

☞ *The numbers refer to the sections of the book.*

THE NON-METALLIC ELEMENTS. — 1. Describe the experiments with muriatic acid. What products are obtained? The test for each? How may they be united again? 2. Describe the experiments with sodium and chlorine water. What are obtained? How is the oxygen tested? If hydrogen be burnt in oxygen, what is the product? What does this show? 3. Describe the experiment with ammonia. What new gas is obtained, and how is it tested? What does ammonia contain? 4. Define compounds and elements. Give examples of each. How many elements? 5. What two classes of elements? What are symbols? What is the symbol of gold? of iron? Explain. 6. What is affinity? What are its three characteristics? Illustrate each. 7. What are molecules? atoms? Why so called? What is said of the atoms of an element? What of the molecules of a compound? What is this theory called? What is the atomic constitution of muriatic acid? of water? of ammonia? 8. What symbols do compounds have? What is indicated by these symbols? Illustrate. 9. What are atomic weights? The atomic weight of H? of O? of N? of Cl? How is the relative weight of the elements in a compound shown by the symbol? 10. Describe the preparation of O from $KClO_3$. What is a reaction? Express this reaction by an equation, and explain in full. 11. Describe experiments with charcoal, sulphur, and phosphorus. What product in each case? How may Fe be burnt in O? What product? What properties of O do these experiments illustrate? 12. What is ozone? Why so called?

How obtained? How tested? 13. What are allotropic states? 14. What are oxides? How named? 15. What are acid oxides? anhydrides? acids? Illustrate. Explain the names and symbols of acids. 16. What are basic oxides? bases? How named? Illustrate. What are alkalies? 17. What are neutral oxides? 18. What is a salt? How named? Illustrate. 19. What are binary compounds? ternary compounds? 20. How is H prepared? Express the reaction in an equation, and explain it. 21. What are the properties of H? Illustrate by experiments. 22. Describe the oxy-hydrogen blowpipe, and the experiments with it. What is the lime light? 23. The properties of H_2O? What is said of its solvent power? 24. What is said of rain water? spring water? mineral waters? hard and soft water? sea water? 25. What is said of H_2O in plants and animals? 26. What is water of crystallization? Efflorescence? Deliquescence? 27. Give some account of the uses of H_2O. 28. How is N prepared? 29. The chief characteristic of N? 30. How is NO prepared? What experiments with it? 31. Define catalysis. 32. What is meant by the nascent state? 33. What is N_2O_3? How prepared? 34. What is NO_2? N_2O_5? HNO_3? 35. The properties of HNO_3? How made? Illustrate by an equation. Explain the action of HNO_3 on metals? 36. What is N_2O? How made? Its properties and uses? 37. What is H_3N? When absorbed by water, what does it form? How is it prepared? 38. What is ammonium? Why is it considered a metal? 39. What is CN? How obtained? For what remarkable? What are some of its compounds? What are radicals? 40. What are the characteristics of N? Illustrate each. 41. Describe the allotropic forms of C. 42. What is CO_2? Its properties? How prepared? Explain. How tested? 43. What is CO? Its preparation? Properties? 44. What are the sources of S? How is it purified? 45. Describe the allotropic forms of S. Its properties? 46. What is SO_2? Properties and uses? Illustrate by experiment. 47. What is H_2SO_4? Ways of making it? Explain in full. 48. What is SO_3? Properties? 49. What is H_2S? Preparation? Properties and uses? 50. What is CS_2? How made? Properties? 51. What elements belong to the O group? Why thus classed? 52. How is Cl prepared? Explain the reaction. 53. Properties of Cl? Explain its bleaching power. 54. Describe the old and new methods of bleaching. How is bleaching powder made? For what else is it used? 55. How is muriatic acid made? Explain the reaction. 56. What are hydracids? oxyacids? 57. What is aqua regia? Why so called? Explain its action on gold. 58. Give the names and symbols of the Cl oxyacids. 59. Give the history and properties of Br. 60. Where is I found? Its properties? 61. Properties of F? How prepared? 62. What is HF? How made? Uses? 63. What elements in the Cl group? What is said of them? 64. Give the history and the sources of P. Its properties? Uses? 65. What is said of the compounds of P? 66. What is said of As? 67. What elements in the N group? Why thus grouped? 68. How is boracic acid obtained? For what is B remarkable? 69. What is SiO_2? What is said of Si? 70. What is said of the C group?

THE METALS. — 71. What are some of the heaviest metals? some of the lightest? What is said of the melting-points of the metals? 72. What are ores? veins, or lodes? 73. What is Fe? Explain the symbol. What three forms of Fe in commerce? How is Fe made fibrous? How does it sometimes lose this texture? Explain welding. 74. Describe the blast-furnace. How is cast-iron made? How converted into wrought iron? Describe the reverberatory furnace. 75. How is steel made? Its properties? Describe the Bessemer process. 76. What is Pb? How obtained? Properties? What is the action of H_2O on Pb? What is said of lead pipes lined with tin? How are lead pipes best protected from water action? 77. What are the chief ores of Cu? Properties of Cu? 78. What are some of the alloys of Cu? Their composition? For what are they remarkable? 79. What is SnO_2? Where found? How is Sn obtained from it? Properties of Sn? Uses? What is said of tin water pipes? 80. What alloys of Sn are mentioned? 81. Chief ores of Zn? How reduced? Properties and uses of Zn? Its alloys? 82. Give the names and symbols of the iron oxides. What is $FeSO_4$, and how is it made? Its uses? What other salts of Fe are mentioned? 83. The names and symbols of the lead oxides? What is $PbCO_3$? How is it made? Explain. What is the chemical name and symbol of chrome yellow? 84. What are the oxides of Cu? What are the chief cupric salts? 85. What oxide of Zn? What salt is mentioned? 86. What is SnO? SnO_2? What is a mordant? What salts of Sn are used as mordants? 87. The chief ore of Hg? How reduced? Properties of Hg? Uses? What are amalgams? Describe the silvering of mirrors. 88. Sources of Ag? Explain cupellation and amalgamation. Properties of Ag? What is said of the compound of Ag and S? 89. Where is Au found? How is it obtained? Properties? 90. Where and in what state is Pt found? How is it obtained from the ore? Its properties and uses? 91. Give the names and symbols of the oxides of Hg. What is calomel? corrosive sublimate? vermilion? 92. What two silver oxides? Their symbols? What is $AgNO_3$? AgCl? What silver salts are used in photography? For what is AgCN used? 93. What is Au_2O? Au_2O_3? What salts of gold are mentioned? Their uses? 94. What is PtO? PtO_2? $PtCl_4$? 95. How was Na discovered? How obtained now? Properties? What is said of its compounds? 96. Where is Mg found? How obtained? Properties and uses? 97. The sources and preparation of Al? Its properties? 98. What is Sb? Its chief ore? How reduced? Properties? Alloys? 99. State the sources, properties, and uses of Bi. How is it now classed? 100. Name the properties and uses of Ni. 101. What is Na_2O? Na_2O_2? HNaO? Its properties and uses? How made? State the reaction. 102. Give the chemical name and symbol of common salt. How is salt obtained from sea water? How are salt lakes formed? What is said of rock-salt? of salt springs? 103. What is Na_2CO_3? Its uses? How was it formerly made? Give the history of the modern method? Describe the process in full. 104. How is $HNaCO_3$ made? Its commercial name? Uses? 105. What is the chemical name and symbol of soda-saltpetre? Its uses? What is Na_2SO_4? Mention other sodic salts. 106. What is MgO? Epsom salts? 107. What oxide has Al? Its sources and uses? What is alum-cake? Give an account of the alums. What is clay? kaolin? 108. What is glass? Describe the four kinds. Give an account of glass-making. What is porcelain, and how is it made? 109. What two oxides of Sb? Its chief salt? 110. What are the Bi oxides? What is $Bi3NO_3$? 111. What two oxides of Ni? 112. How was K discovered? How is it now prepared? Its properties? What is pearlash, and how is it made? What are the chief potassic compounds? What is said of gunpowder? of $KClO_3$? 113. Where does Ca occur? How is the metal obtained? What is CaO? Its preparation, properties, and uses? What is mortar? hydraulic cement? What is said of $CaCO_3$? of $CaSO_4$? of $CaCl_2$? 114. Give an account of Sr and its compounds. 115. What is said of Ba and its compounds? 116. The chief compounds of Cr and their uses? 117. Describe Co. What compounds are mentioned? 118. Give an account of Mn. What oxides has it? Which is the most important? Its uses? 119–122. What is said of Cd? of Ir? of W? of U?

CHEMISTRY OF THE ATMOSPHERE. — 123. Prove that the air contains free O. What are its other constituents? Is it a compound of these gases? 124. What are the products of the burning of a candle? Prove this. 125. Prove that the candle takes O from the air. 126. What is all ordinary combustion? 127. Why is a draft needed in a stove? 128. What is a combustible? A supporter of combustion? What is true of these terms? Illustrate by an experiment. 129. What is necessary to make O unite with a combustible? 130. Show that combustion is self-sustaining. 131. Show that the point of ignition is different for different bodies. 132. When are products of burning gaseous, and when solid? 133. What is true of Mg? Give other examples of the kind. 134. Show that O is not the only supporter of combustion. 135. What is true of the earth's crust? 136. Show that the materials of the earth are the results of burning. 137. What three states of O? What is said of the third? 138. What is decay? Describe the process, and explain in full. 139. Explain the causes of decay. 140. What is rusting? Show that it develops heat. 141. Define and illustrate spontaneous combustion. 142. What is said of decay in animals? What three kinds of food? What does milk contain? What does bread? What purpose does each kind of food serve in the body? How is the body heated? Describe and explain in full. 143. What experiment illustrates the burning of sugar? 144. Give facts concerning the daily consumption of O. 145. Describe the formation of the embryo plant in the seed. 146. Give an account of the growth and structure of the plantlet. 147. What are organic beings? What is organic structure? 148. Explain how plants get their food. 149. Of what does the inorganic portion of the plant consist? Why are manures necessary in agriculture? Why are fields allowed to lie fallow? Explain the principle of the alternation of crops. 150. Give an account of the organic portion of the plant. 151–153. How does the plant get H and O? C? N? 154. Show that the growth of plants is a chemical process, and how it is carried on. 155. Show that plants purify the air for animals to breathe. 156. Show that plants prepare food for animals. 157. What is said of the relations of plants and animals? 158. Give a full account of starch. 159. What is sugar? What two kinds? 160. What is the symbol for cane-sugar? From what is it obtained? Describe its manufacture from the cane, and the refining process. Give some facts in the history of sugar. 161. What names has $C_{12}H_{24}O_{12}$? In what is it found? How may it be made? Its uses? 162. What are the sources and uses of dextrine? 163. What is the symbol of cellulose? How is gun-cotton made? What is said of it? What is collodion? Its uses? 164. Describe the manufacture of vegetable parchment. Its uses? 165. What is said of $C_4H_6O_6$? of its compounds? What is $C_6H_8O_7$? $C_4H_6O_5$? C_2O_3? $H_2C_2O_4$? Give an account of tannic acid and tannin. 166. What is the composition of fats and oils? What two classes of oils? Give some account of both. 167. What is soap? How do hard and soft soap differ? The composition and uses of glycerine? 168. What is said of wax? 169. How do gums and resins differ? Mention the chief examples of each. 170. Give some account of vegetable alkaloids. 171. Describe the proteine bodies? Why are they important?

DESTRUCTIVE DISTILLATION AND ITS PRODUCTS. — 172. What is destructive distillation? 173. Describe the preparation of charcoal. 174. What are the products of the distillation of wood? 175. Describe wood naphtha. 176. What is creosote? Its uses? 177. Give an account of paraffine. 178. What is asphalt, and for what is it used? 179. What is slow destructive distillation? 180. Explain the formation of mineral coal. 181. What two kinds of coal? How do they differ? 182. How do the products of the distillation of soft coal and of wood differ? 183. What is found in coal-tar? What are the uses of carbolic acid? Of aniline and its compounds? 184–186. What is said of benzole and nitro-benzole? of toluole and cumole? of naphthaline? 187. Give the history of the coal-oil manufacture. 188. Explain the probable origin of petroleum. 189. How is petroleum obtained? 190. Give the history of coal-gas. 191. What is flame? 192. What are the two conditions of illumination? 193. Explain the burning of coal-gas. 194. Describe and explain Bunsen's lamp. 195. What is the best shape for a gas-flame? 196. Describe the Argand burner. 197. Explain the Bude light. 198. Show that a burning candle is a miniature gas-factory. 199. Analyze the candle-flame.

FERMENTATION. — 200. State the nature and causes of fermentation. Explain apparent cases of fermentation without a ferment. What two kinds of fermentation? 201. Explain the alcoholic fermentation. How is it brought about in grain? 202. Explain the formation of diastase in plants. 203. Describe in full the process of brewing. 204. Describe the distilling and rectifying of spirits. 205. What is ether? Its properties and uses? Name other ethers. 206. Explain the acetous fermentation. Describe methods of making vinegar. 207. Describe and explain bread-making. How may bread be made without fermentation?

INDEX.

☞ For concise statements of the leading topics of the book, see the SUMMARIES, which will be readily found by means of the *Table of Contents*. The references in this Index are only to the fuller treatment of subjects in the body of the work.

A.

B.

O.

P.

Q.

R.

S.

Cambridge: Electrotyped and Printed by John Wilson and Son.

APPARATUS.

The articles in the List on pages 181–184 can be furnished at the following prices: —

No. 2, (cylinders and ground-glass plates), ...	$2.50
No. 3, (jars and ground-glass plates), ...	3.00
No. 4, ...	.70
No. 5, ...	4.00
No. 6, ...	1.00
No. 7, ...	1.00
No. 8, ...	2.00
No. 9, Bunsen's lamp, $1.50; alcohol lamp, ...	1.00
No. 10, ...	5.00
No. 11, ...	5.00
No. 12, ...	1.50
No. 13, ...	.50
No. 14, ...	1 75
No. 15, ...	3.00
No. 16, ...	.50
No. 18, ...	2.50
No. 19, ...	.50
No. 20, ...	.50
No. 21, ...	.63
No. 22, Half a pound of tubing, $\frac{1}{4}$ and $\frac{1}{8}$ of an inch in diameter, ...	.30
No. 23, ...	.80
Two 25-gallon gas-bags, with stop-cocks and tubing, . .	35.00
An oxy-hydrogen jet, ...	18.00
A copper flask for making oxygen, ...	7.00

For further information, please address

MR. SAMUEL F. HUNT,
CAMBRIDGEPORT, MASS.

www.ingramcontent.com/pod-product-compliance
Lightning Source LLC
La Vergne TN
LVHW011207110826
845150LV00006B/1343

* 9 7 8 1 4 2 5 5 1 8 3 6 3 *